PAST, PRESENT AND FUTURE AS TIME IN THE AGE OF SCIENCE

SECOND EDITION

SAMUEL K. K. BLANKSON

WORKS BY SAMUEL K. K. BLANKSON

The metaphysical Foundation for Physics;
Why time is not a natural phenomenon;
The Mathematical theory of time;
How old is the universe by our time;
The Einstein theory of space-time without mathematics;
Dead End (a novel);
Time in Science and Life---The greatest legacy of Albert Einstein;
How religious scientists play down the greatest of Einstein's achievements;
The coming revolution in physics;
Time and the Application of time;
The Logic of time in the universe;
Philosophical Essays;
Past, present and future as time in the age of science

First published in Great Britain in 2015 by
PRACTICAL BOOKS

Published by Blankson Enterprises Limited
4 Poplar House, London
SE4 1NE
www.practicalbooks.org

Copyright © 2015 by Samuel K. K. Blankson

All rights reserved. No reproduction, copy or transmissions of this publication may be made without written permission. No paragraph may be reproduced, copied or transmitted save with written permission or in accordance with the provisions of the Copyright Act 1956 (as amended). Any person who does any unauthorised act in relation to this publication will be liable to criminal prosecution and civil claims for damages.

ISBN 978-1-326-53586-5 – Paperback
ISBN 978-1-326-53591-9 – Hardback
ISBN 978-1-326-53589-6 – E-Book

A CIP catalogue record for this book is available from the British Library.

CONTENTS

PREFACE 5

INTRODUCTION 17

CHAPTER ONE 25

    CLARIFICATIONS 25
    WHY DURATION IS THE ENGINE OF TIME---THE CAUSE OF THE
    SENSE OF TIME 33
    WHY SECULAR TIME IS NECESSARILY DISCRETE 37
    WHAT A MOMENT MEANS 41
    THE IMPORTANCE OF PHILOSOPHY 50

CHAPTER TWO 55

CHAPTER THREE 60

    THE MERGING OF SPACE WITH TIME 67

CHAPTER FOUR 80

    THE FOUR AGENTS THAT CAUSE WHAT WE EXPERIENCE AS
    TIME. 80
    THE ORDER OF TIME SEEN AS A MATTER OF ARITHMETIC 90

CHAPTER FIVE 105

    TIME AND CONDITIONING IN THE UNIVERSE 105

CHAPTER SIX 113

    THE NATURE OF TIME BEFORE AND AFTER EINSTEIN 113
    SECTION 6 (A) - TIME BEFORE EINSTEIN 113
    SECTION 6 (B) - WHAT IS SECULAR TIME? 123
    SECTION 6 (C) - TIME AFTER EINSTEIN 125

CHAPTER SEVEN 132

    ENTROPY, GRAVITY AND TIME 132
    ENTROPY AND TIME 8 144

GRAVITY AND TIME 9 149
APPENDIX I: 153
  TIME AND QUANTIFIED TIME OR THE PASSAGE OF TIME 153
  WHAT IS MEASURED BY THE CLOCK? 161
APPENDIX---II 166
  THE PRINCIPLE OF MATHEMATICAL EQUIVALENCE 166
APPENDIX---III 171
  WHY SPACE ON ITS OWN IS NOT "SPACE-TIME" 171
APPENDIX IV 177
  THE MISCONCEPTIONS OF TIME IN RELATIVITY 177
CONCLUSION 188
REFERENCES 203
INDEX 207

PREFACE

After fifty years of researching time in all of its aspects, I have come to the conclusion that no matter what logic is produced and by whomsoever, there will always be people including scientists who will argue for the religious view of time, all because people simply do not want to accept that death is the total end of human life. A large part of human knowledge is distorted or corrupted with human emotional beliefs. Yet, still, it came as a complete surprise to me that time is the most contentious subject under the sun passionately debated entirely in religious terms.

Nevertheless I believe that all discussions of time should begin with its clear and objective definition because although we all use something we call time but define its nature and provenance differently due to our religious beliefs, leading to contradictory theories about nature and reality. Another reason is that science demands an objective definition it can use. For my part, I accept that after centuries of thinking about time logically, we've come to the conclusion that it is constructed by man as relation between points (Bertrand Russell); that it is also limited to a frame, so "there are as many times as there are frames", and cosmic or divine time cannot exist because the Lorentz discovery of local time means anybody can begin his or her own time from anywhere (Albert Einstein); and, therefore, time is metaphysically caused by the instantaneous spread, visual experience or any contact with the outside world (Professor A.N Whitehead.) Also, from the scientific point of view, Professor Sir Arthur Eddington (the founder of astrophysics), says that any contrary notion of time amounts to making "meaningless noises" as time is no longer seen as running smoothly ('even-flowing') through nature, and that, since this is the logical view, we humans have no other sensible or cogent explanation for time---and, quite rightly, he credited it to Albert Einstein.

Therefore, as mere followers and supporters of these great thinkers, altogether, we now see time as the artificially generated periods by means of regular or repetitive motions for the total regulation of all activities and life generally, since it is 'constructed' from the general experience of the physical universe multiplied by motion or mathematically divided with points. Thus sentience, arithmetic, the ability to count and a theory of numbers were required. This shows that it was not an easy thing to acquire, and arose after man came down from the trees. No wonder it has exercised human ingenuity for centuries, not least about the passage of time.

However, from the scientific point of view, the passage of time is the least troublesome; we can even tap the finger to indicate the passage of time without arrows. For time itself does not move; only *the cycles or motions we use to reckon time do move and deceptively seem to us as the passage of time, which, being arithmetical, can only advance through numbers---the years, for instance. The motions we use to reckon time are repetitive, which shows that it is not the motion of time at all. Time is 'Being'(existence); and we sentient beings while 'existing' count repetitive cycles or motions to indicate the passage of time passing by our 'Beings', or how many cycles have passed by. It is a very complex and mysterious matter but human in origin. Take the years, for example: there are no 'years' in nature. There is only one year repeated to be years all the way to the centuries which are "**by-passing**", say, the Alps, year after year after year. It's not the Alps that are moving to provide the years but the cycles or repetitive motions we use to mark time. Thus any repetitive motions can be used to the same effect---tapping the figure is the same thing. Otherwise time does not exist as a physical entity. It's a complement to the human mind. That is why Russell deduced that we 'construct' our time---well, in the absence of a universal time our time had to come from somewhere, and we know that we are the ones who count the years to determine age. The factors for use in the construction of time may exist throughout the cosmos, but the time is constructed by us and does not, certainly, cannot be something running all through the cosmos and the same everywhere. Whatever 'Beings' are there in the cosmos would have to construct their own time systems by similar logical methods--- this last point is my own personal belief as a certainty based on logical reasoning. The problem with logic is that it is science, meaning what we can trace or know in nature as of substance and/or really in existence in the cosmos. Like shadows, aura and* *influence, they may not be substantial, but they do certainly exist, or do occur. Einstein often referred to 'logical thought', or 'scientific thought' and he was right. He and Aristotle (and probably Russell as well) were the most logical thinkers ever to grace this planet. Discoveries, inventions, literature, art and scholarship are necessary to create civilizations; but logical thinkers constitute the foundation of science.*

As stated above, the culturally necessary units of time in the various lengths arise from the division of the earth's motions (and space traversed) by points into units or intervals, whereby the closely placed points provide short units and distant points give us long units of time like the year. So the seconds and years are the best confirmations for this theory. Time's divine status, arrows for its passage and absolute nature can all be consigned to the rubbish tip.

As defined in secular terms, time is now a branch of astrophysics; for so long as the year is our basic unit of time from which all other units are derived as fractions, time can be consistently deduced from experience: in any part of the universe any regular motions divided by points as required (into short or long intervals) will provide time units for cultural use as the logic of time in the universe. Time has no existence outside the human mind and what does not exist cannot be passing by. It is purely psychological counting of physical cycles as time units like the physical orbits of the sun as years.

One orbit of the sun is 'a year', a unit of time out of which all other units are derived as fractions thus: time in culture is known only in units. We pare down a year to the seconds and even lower. Each second is equivalent to about twenty miles in physical distance round the sun. At this rate, even a millionth of a second is quite a definite amount of space; so all of our time is derived from the year or the earth's orbit of the sun physically, which confirms the theory or notion that time is the equivalent of the secular coverage of physical distance, by which we are able to tell the difference between the lengths (or duration) of specific time units----that the hour is longer than a minute and so forth. Nothing is left of time to be explained with theoretical postulates as the universities, especially CUP, has been foisting on mankind to no avail. They still insist that Past, Present and Future cannot be wished away, and that even Einstein could not explain them to anybody's satisfaction. Yet he did. In logical thought, Past, Present and Future are truly illusions. The past is with us; it never went anywhere. It is what we carried with us to the present: our clothes, houses and other possessions from yesterday are what we are using today! For, after all, there are no days at all in nature. There is only one day. The temporary blips of the earth's revolutions across the sun do not change sunshine from day to day. The daylight, bar clouds, is on from the sun all the time. The earth's revolutions create the illusion of successive days, otherwise, from the sun's point of view, they do not exist. Thus the past does not still exist anywhere to be revisited by those religious theories about time travel.

The present is the past brought with us (we're still living in the same houses, for instance), and the future is unknown---which is what primitive man told us but we would not listen. It's now established beyond doubt, at least in my experience, that people just happen to be welded to ideas of reincarnation. To my mind it's as simple as that; and basic to that belief is that time is so mysterious that we can never know what it is. In fact, it is as defined above. Of course it is very closely associated with life and it is conceded that nobody knows why there is life; but time and life are two different things: time requires points, meaning that we had to come to be before knowing how to 'construct' out time. The long and short of which is that as the past does not persist anywhere to be revisited in the past, the future, too, is nowhere to be reached ahead of time (the time has not even been constructed yet); so the future is nowhere until it arrives through the earth's motions round the sun---and we know that it may never arrive: When people die (and we die all the time) their future is lost; it will never arrive, for everything is what it is only in human perspectives. That is why we cannot use animals, say, lions to do our thinking for us. While living as human beings we can of course use the imagination to speculate about the future; yet the imagination on its own is not reality, and yet outside reality there is only chaos not truth.

But we need to know the truth. If you take a million people with the freedom to speculate, you'll get more than a million viewpoints, as some people would have more than one opinion---we need truth to counteract that ruinous tendency in mankind. Man is powerful and at the same time highly vulnerable. Altogether, time is important and should be defined properly in logic. The best way to do so is, first to abandon its divine suppositions, which we can now undertake reasonable confidently, due to the influence of Albert Einstein, as Sir Arthur Eddington has stated above in his definition of time. And it's about time people, including all those religious scientists, realised that Einstein's theory of time is far more important than all his theories put together, for after all we live by *time, and we have to live. The strangest thing in the world, for me, is that his theory of time came out of a simple remark that the Lorentz local time idea can be defined as time 'pure and simple'.*

As if that was not shocking enough, the greatest philosopher in the world at the time, our very lovable own Bertrand Russell, asked the most important question ever posed about time by asking what then is mechanised in the clock. I have since in my own little way been trying to answer Russell's query as detailed in this book. I think that ultimately time comes from nowhere and gets mechanised in the clock as nothing other than our habit of counting cycles and calling them 'time units' based on the period inherent in the earth's orbit of the sun pared down to the seconds. Everything in time is based on the earth's motions, for time is action and how long any action takes. Let me explain. Action is caused by energy, not time; so action in itself is not time or time controlled; it'd be either spontaneous or chemical; but time is then inferred from the action or contact. The time (or time normally), arise because of action, as we want to know the duration of any action; it is after the discovery of how to mark time that we can control activities by means of time for the purposes of creating civilization. So time was invented out of the conditions in the natural world and does not exist in the cosmos as a natural phenomenon.

Action takes time, meaning any action has to cover a period; this period is called 'duration', that is to say, 'during the period of the action'. But originally the concept of time was inferred from action---by counting cycles and calling them the number of cycles that any action took.[1] That is the mechanism the clock was invented

---

[1] Duration itself is caused, as explained below in the book. What causes duration is what

to record----and can now be used to cover action through prediction and action after the event, meaning planning purposes. All this has been explained in the book; without activity the need for the sense of time does not arise, reasoning and intelligence are also connected to the sense of time;[2] hence even animals do have the sense of time, albeit without the linguistic, mathematical and mechanical skills to mechanise it in the clock. By adding language, religion, fear of death, cultural practices, psychology and so forth to duration between events, time has become the most intricate subject on earth. But in the beginning although all the parameters for constructing time were there, man had no notion of mechanised time when he came down from the trees. As Russell has observed, time was a construction.

Finally, let me try to explain Professor Whitehead's highbrow or metaphysical definition of time---as I understand it (it's always a fallible attempt, for no one can claim to know anything for sure about time. All he or she has to do is to make any attempt logically valid). What he means, I think, is that time is an activity, vision or contact of any kind in nature; hence daylight can be regarded as the most basic form of human interaction with nature. Man's first attempt to reckon time was by merely marking the days as they passed and calling the process 'the passage of time' (I saw my grandmother doing that in the 1930s.) We still say the passage of

---

we experience as time, and, although secular (mostly chemical), it may not be knowable in all cases. Thus aspects of time will remain mysterious forever.

[2] Existence without the sense of time is dour. Creativity, thought and civilization are all the consequences of having the sense of time. The reason 'Being' or life is so closely associated with time that they cannot be separated logically. The curse of religion prevented any thinker from discovering this idea of secular time until Einstein. Then God said 'let there be Russell and Whitehead to support him' and all was light!

time is exemplified by the passage of the days and nights---but there are no days; there is only one day, and it does not pass.³ That is the quandary. Like Russell, there is no doubt that Professor Whitehead was extremely clever. He was also the first to note that the world of sense is a construction not an inference in so far as we can only see by means of light, yet light is also matter---in fact, the smallest bits of matter in our universe. So it means we construct images with the basic materials in existence. For man reality is limited to what light can reveal to him.

That is the definition of reality under the quantum theory---we construct reality with quanta, not infer it through the use of the mind, nor even perceive the world on the principles of Plato's Idealism. Due to their professional vanity and inherent disrespect for philosophy, scientists shared the Nobel Prize for QED without mentioning Whitehead's theory, yet QED is the physical evidence that his suggestion was true---what we regard as reality of the external world is actually inferred from physical elements absorbed from the interactions between photons and electrons as Richard Feynman has shown. For once Idealism was logically refuted by G.E. Moore, there remained only one route for any meaningful philosophical reasoning. That's empiricism, but that doctrine ends

---

³ Let me stress again that time does not pass because it does not move; what is moving is the repetitive cycle we count as the rates of passing time. It is an irony or a mystery, a paradox or sheer ignorance. Sunshine as daylight (or 'the day') neither moves nor passes by; it is on constantly. Rather we count the motions of the earth across the sun as 'days' all the way to the years, and pare the year down to our seconds and all other units of time. This is very important in the discussions about time. The main reason I am frustrated is that nobody understands this, yet it is true---there is only one day and it does not move. The concept of 'the passage of time' is completely mistaken. It occurs in culture but not in nature. All history (even the difference between yesterday and today), is the passage of events not time---we rely on memory and records.

in scientific enquiries and discoveries and is based ultimately on quantum electrodynamics, known as QED. Thus the Nobel for QED should properly have been shared between Whitehead, Russell and Feynman: Whitehead for the original theory, Russell for popularising the idea in his book Our Knowledge of the External World linking the process to physics and Feynman for proving it so ingeniously in physics.[4]

After all, this theory, believe it or not, is the end of all human intellectual quests. The notion that light is reality, like Plato's Theory of Forms, proposes a solution of the ultimate query about human existence. It is the beginning and end of all philosophy and science, being and consciousness. But whereas Plato's theory has been refuted as not true of the world, reality as constructed with light energy is as solidly true as there is light, the same light by which we see the world. So in the end, the truth came as a very simple notion: what you see is what light shows there is, and the scholars who discovered this idea (Einstein, Russell, Whitehead and Feynman) deserve the highest honour. Individual academic, literary or cultural honours are not enough. We need a huge monument somewhere accessible so that people would ask 'what did they achieve?' and through that learn the greatest philosophical truth that settled the final conundrum about reality.

The next quandary to confront this very clever American man--- Professor Whitehead[5]--- was the problem of time because, as I said

---

[4] The other scientific contributors were minor figures, but the Nobel committee seems to think that it is important to spread the awards, thus bringing nationalistic tendencies to bear, which I believe will eventually make them worthless. For instance, even though Einstein had been rewarded already, he and Bose deserved another. The awards should just follow merit.

above, they are linked. He realised that to account for secular time under relativity some event (action, inaction or vision) has to occur; then counting cycles or applying units of repetitive cycles would give the time or duration of whatever had occurred.[6] Convoluted, perhaps? Yes it is. That's what time is, and the reason it's so mysterious---much easier to assume that God bestowed it on man and forget about the rest! By this theory Plato was not that clever, only a good writer; for Idealism makes thinking easy, but nothing in life is that easy, philosophically. Under the quantum theory, all knowledge cannot be mental because the light from anything can be blocked from reaching the human eye to result in seeing or knowing anything. Thus nothing without its light emissions can be seen or known. This is the final proof that all knowledge is not mental. Reality is out there and its lights must reach man for it to be seen or known. Idealism is dead, that is why the religion based on it is also dying---though only slowly, but it'll go.

For the logician the problem of time is to find an explanation or theory for the long period of the day or night, and that's when the counting of cycles came in. Any cycle will do, even tapping the figure is quite alright; but through human ingenuity we have evolved mathematical techniques for paring the yearly cycle down to the seconds. So we say each episode lasts for many hours, or so many units of time.

---

[5] It was bound to happen that these four thinkers mentioned above would have something interesting to say about time, too.

[6] The sense of time is inferred from events---or even inaction.

INTRODUCTION[7]

It's sheer human arrogance and groundless vanity to speak of Man and the cosmos in any discourse----we are so infinitesimally insignificant and unsure of anything at all; the nearest equivalent in vanity and mystery is like a new-born baby telling the parents what to do---how did it get the knowledge or how did it know the language. It'd be a mystery. So is breathing mankind in the world of lifeless matter, except that we have the brain. With the brain we are able to probe nature all the way to the billions of galaxies to infinity. So what is the status of this frail and easily-destroyed man and his brain?

Another mystery is that everything begins with time; everything has to be in time or does not have the credentials of existence. Relativity, too, is relevant because it started the inquiries about the nature of time; before that we just took it for granted as fixed by God, it is absolute, generally covering the whole universe and the same everywhere, as Professor Sir Arthur Eddington put it (and will be quoted many times throughout this book because many

---

[7] This book is the second edition of one published in 2013, but the Introduction is completely new as the original has been discarded.

scholars just cannot even begin to accept that that is the truth about time today.)[8] Eddington wrote: "Prior to Einstein's researches no doubt was entertained that there existed a 'true even-flowing time' which was unique and universal...those who still insist on the existence of a unique 'true time' generally rely on the possibility that the resources of experiment are not yet exhausted and that some a discriminating test may be found. But the off-chance that a future generation may discover a significance in our utterances is scarcely an excuse for making *meaningless noises"* (The Mathematical Theory of Relativity, ch.1.1.)[9]

But despite the title of this little book, the quantum and QED are mentioned only briefly here and there, though the trust of the argument is directly related to the relativity notion of time introduced by Albert Einstein. And although this is not a book on physics as such, the idea that time is seen differently in science is shown to be completely mistaken. There is only one system of time in the world---it is the same one that scientists use in all their researches. Einstein changed time for all mankind not that he invented a new time system for use in physics or relativity, far from it. His suggestions led the philosophers to declare that time is secular and 'constructed' by man as will be explained in this book, and the quantum comes in because it is based on time; it materialises from a fraction of the second, and the second is also a fraction of the earth-year, while the year is our basic unit of time out of which all other units are derived as fractions. So a new logic

---

[8] I have great difficulty getting any publisher merely to read my manuscripts because they usually reply that the problem of time was settled by Newton, so they'd not read my books even before rejecting them;. Some 'divine institutions' like Harvard, Princeton and the Royal Society would not reply to my correspondence at all.

[9] The italics are mine. For this book is based on his true, honest and categorical statement and Russell's judgement that time is a construction, and that universal time is abolished---which means we have to research how we obtained out time, and why the divine institutions would not even reply to my communications because I stress these truths beats my understanding.

has been introduced and it makes time a) secular; b) constructed by man; and c) necessarily discrete.

All these dire inferences began (as will be made clear in the book) when the great Bertrand Russell deduced that according to the Einstein notion of time, time does not run through nature but is constructed by us and therefore by implication necessarily discrete. So cosmic time is abolished not only in science but for everybody else; it's been quietly ignored or neglected as an intellectual subject for debate because people----including the scientists themselves---just cannot imagine that time is no longer fixed by god, no longer absolute, and does not run from the distant past to the present and the infinite future, and that it is just is. That it's there even before we are born and leave it behind when we die, and so forth; all of which may be described as 'The Traditional Notion of Time'. Anyway this is what I have discussed in the book.

Thus it is not a technical book on philosophy. It is rather something like "a general consensus treatise" about the ordinary thing we all know as 'time'---time for sports, for travel, for work and for doing everything we do.[10]

The sense of time, the entire human idea of time, is based on the passing days and nights. To know or understand time we learn that it is passing in days one after another, or nights; it is the same thing. This idea is then translated into physics, memory and misuse of the human imagination, metaphysics, religion, mythology, mathematics and ordinary linguistic practices with theories, customs and attitudes to match. No wonder time has become the most mysterious subject in the universe. We read time into everything. So when Einstein said it is limited to a frame nobody could understand him; when Russell also asked how we get it into

---

[10] Cambridge University Press (another divine institution) wrote to reject my book because it is addressed to the general public but their philosophy list does not serve that market. On the other hand they have published (or foisted on mankind for its sins) a hefty tome about the arrow of time which does not even know that these assertions by Eddington and Russell do exist.

the clock, he was ignored, and Whitehead's bemusing theory that it is instantaneous and therefore by implication cannot move, was linked to Russell as 'philosophers preaching their usual incomprehensible abstraction unrelated to reality'.

In fact, they are all right---precisely as I have defined time above which, of course, means that what I write, to the vain literary gurus and academic morons, is not even worth looking at. Nobody has ever asked to see my manuscripts. They'd call for a summary and dismiss the work out of hand---some would even go as far as demeaning themselves by making rude remarks; and another begged me to go to the academic agents because not being a mathematician he could not understand a word I was saying! Yet I write in the plainest possible English language because I want to be understood and accepted----no chance. They retort that I am doing so merely in the hope of increasing sales. Cambridge replied that my material is too short, and, also, they did not like to publish philosophy written in the language of the ordinary man. Oxford was wise enough not to pass judgement, but merely said in their rejection slip that they would make no comment about the philosophical and scientific aspects of my work----well they might, because they would not understand it anyway. It's the gibberish from Wittgenstein they recognise as 'great philosophy'. That Russell snubbed him spectacularly in The History of Western Philosophy because he was trying to destroy physics is never mentioned. Thank God nobody can destroy physics, and just for trying should be ignored as if he had never existed. MIT, Harvard, Princeton and The New Scientist did not bother to reply at all to my submissions. On the other hand, NATURE wrote me a very helpful reply; no wonder it is regarded as the best in the scientific world. I have heard of new ideas being initially overlooked, but I am basing my ideas on those of Eddington, Russell, Einstein and Professor Whitehead. Harvard should have known what that means, and what about Einstein's old institution, Princeton?

Anyway, the passing days is not (does not constitute) passing time because there are no days in the universe. There is only one day. Man came to be and lives by the merciful bounty of astronomy, and in astronomy there are no days----there is only the sun shining constantly. It is what we know as 'Daylight', or 'Day'. The night results from the revolutions of the earth blocking the sunlight for a period---hence The Day and Night system which has given us the sense of time 'passing' by. Even worse, there are no years either. Yet the concept of passing time has spawned the most intricate of metaphysics, religion, philosophy and all the mysteries about time. However if we try (very hard) to logically analyse time, as Professor Whitehead observed, we find that time, as daylight, for instance, does not move----only the cycles we use to mark the passage of time do move. For example, when the sun is on, we count cycles and say it is on for twelve hours, or twelve cycles. How we create the units of time by using repetitive motions have all been explained in the book, but I don't expect the 'great scholars' of Oxbridge to understand it because they are too great---- the mere mention of Oxford or Cambridge makes them infinitely invincible world-class scholars with whom black savages like me from the old British colonial rain forest should not dare to over-reach themselves by trying to engage in discourse. Usually when they are about to accept my work they ask for C.V. and for me, as soon as I send my C.V. the rejection slips would be issued.

Well, the truth is that in the popular imagination and folklore, yesterday is the past, today is the present day, and tomorrow is the future. That is what all the religions and metaphysics and traditions are based on. From that we got the idea that the days and time as a whole are passing by. Yet there is only one day in astronomy! I am not the one to blame. There simply is no universal time, as Russell pointed out. I have thought about time without the obvious mysteries, because I think they are all false once time is seen as originating from, and limited to, this planet. That is what this book is all about; but there are defects in the book just because nobody would give me an ounce of literary assistance. It may well be they do not understand the mathematics by which it is presented!

This book is not only aimed at the lay reader. The word 'lay' would make it an intelligent book for intellectuals who merely happen to be non-professional thinkers, the kind of people commenting on every subject in the letters columns of the serious newspapers. Alas, they are not the only readers. Everybody reads something and is therefore a book for everybody, since it is about time and time is exactly like life----we all have it, know it and use it. Nobody can do without time.

Before the Lorentz-Einstein notion of time, everybody relied on Newtonian time---general, absolute (meaning it does not change, fixed by divine power), covering the whole universe and the same everywhere. The general idea was that it just is. The new theory is that it is none of these eternal things as explained in the definition above----there are as many times as there are planets, thus making everything that is said about time in the religions totally untrue. And the new theory is scientific because it is derived from objective experiments.

The reaction of the world as a whole has been to ignore the new theory as if it has never been put forward. All this has been discussed in the book for that is where I come in, also to be ignored by the world as if I had never existed. One condition was that the new theory has to be in accord with astronomy, and I have shown that it is, because there is only one day in astronomy and that is what my theory is based on. The sun shines on us as daylight or the 'day'. Only one day; it does not go to sleep as we do during the night time, which is caused by the earth's revolutions. It is a pity the so called great academic institutions of this world cannot understand this simple idea and continue to produce mighty tomes about time rather as if they're illiterate, praising Einstein for his theory of gravity, when, in fact, his theory of time was his greatest intellectual achievement, even Bertrand Russell---the modern world's Aristotle---thought so.

PART TWO of the book is short because it consists entirely of Appendixes (as thought expansions) culled from previous publications that I have used and re-used in many other books. I am very old (born 1938) and very ill and frail and get no assistance from any quarters. My son helps me as my literary agent, editor and publisher, but he is only an engineer, although a saintly genius, with about 20 books to his credit. However I cannot worry him too much as he already works too hard. When you see a full-page feature article in The Guardian about any writer, you tend to believe that the Nobel Committee's urgent call is not far away![11] Thus I re-use many of my old essays in subsequent books when I am too ill to write a whole one.

I agree that this is a literary crime, and for that I apologise to the reader. I believe many of the purely printing and literary errors have been corrected in this second edition as far as possible. But if I were under an obligation to provide a defence, I'd plead that I normally used published material to make up a new book just to bring out a new theory that could not otherwise get published. In this edition, for instance, the definition, the Introduction and Section One contain many new ideas to justify this publication as a separate book. For I think all the time, even in my sleep. Thinking about time takes all my life, it absorbs me completely, day and night, though it's the most unrewarding use of the human intellect. The reason is that life and time are bound up together into one, and life remains mysterious and inexplicable; people believe time too to be as inexplicable and mysterious as life, claiming that time has infinite past and infinite future, and they have all built their religions on this idea of time. On the other hand, this book and the contents of my nine other books on time accept the scientific evidence (according to Professor Eddington provided by Einstein's

---
[11] See "BUSINESS SENSE ON DEMAND", Guardian, 23rd Feb. 2007, p.5.

researches), that time does not flow through all nature and the same everywhere. Russell also said we construct our time. So it means the days do not exist naturally in nature. In astronomy there is only one day as I keep reminding the reader, for it is the basis of my theory of time. Man creates the days and weeks and months and years all the way to the millennia. To me this is exciting stuff. It tells me a lot about the nature of life. No wonder Professor Eddington castigated all those who think otherwise. I agree that it gives us a new meaning to human life on earth whether we like it or not for religious reasons. So if my ideas appeal to any reader then he or she should kindly forgive my minor indiscretions because I am compelled to abide by the scientific evidence---nobody should try to contradict physics due to religious beliefs.

KWASI BLANKSON
LONDON 2015

CHAPTER ONE

CLARIFICATIONS

Publishers do not know everything; they're also human and make mistakes. But they also constitute the only effective avenue for publicising the ideas of the rare geniuses who carry civilisation on their shoulders. The list is endless both ways---for good and bad decisions. Presently they all seem to be yearning for only dramatic theories of time.[12] But there does not appear to be any such magic coming from anywhere. There have been changes since Einstein as I will try to demonstrate in this book. However the irony is that time remains the same for everybody. **The logical or rational solution of the conundrum is to find an explanation that is logically consistent with time as we know it without any trace of mysticism and religion, which is now possible.** But so long as time from all angles remains the same (leaving the days and hours

---

[12] 'Time changes with speed', 'gravity slows time', 'time travel is a scientific possibility' 'you could meet your grannies before they're married' and so forth. Otherwise they don't want to know. Yet if cosmic time is abolished so that time becomes secular and therefore discrete, then how could it perform any of these miracles---without God?

and minutes as they always are), the logical approach, being difficult, gets no hearing at all only mockery. It hurts. I have already published a small book predicting that physics will one day have to change course due to the theory of time---and that, I believe, will definitely come to pass.

Meanwhile, the Cover Story of the New Scientist issue of 15 June 2013, entitled "Space versus Time..."[13] caught my attention whilst in the course of writing my little book about Post Relativity time.[14] "And what is that monster?" you may ask, and you could be right. Many scholars avoid this Nobel Prize subject perhaps because it is difficult. I am not even sure I have done it justice.[15]

It is the view, originating from the Dutch Physicist H.A. Lorentz and adapted by Einstein as the basis of his theory of frames, I think, and amounting to the philosophy that time can begin from anywhere and therefore there are as many times as there are inertial bodies or frames. Strange as it may sound, that single scientific discovery forms the basis of the new definition of time which sees it as discrete and not, definitely not, running smoothly all through the universe from the distant past to the infinite future, since, logically, 'the distant past' is owed to memory as history, and 'the infinite future' is fantasy in astronomy.

---

[13] Of late many academics are writing about serious philosophical quandaries as though they're minor issues in science to which anybody can make a contribution. Yet philosophy is unique and remains extremely important---not a field for small contributions but a rounded view of nature.

[14] People still believe that time is just is, not that there is now pre and post Einstein notions of time, yet Russell and Eddington stressed this all their lives. I am not surprised that I am ignored for even they got nowhere. Human beings are not good otherwise politics won't be so messy. The only man to know more about human nature was Freud. In an area of high intellects, high stakes and phenomenal rewards, nobody would admit that you have said anything memorable---they would rather kill you and steal your Nobel!

[15] This piece was originally written as a journal paper so many of the references overlap with other notes in the book.

Thus universal, divine, absolute and fixed time for the universe are all abolished under relativity[16]---a very serious intellectual challenge requiring the redefinition of almost everything, including religion, the Day of Judgement and life itself because without time there can be no life, since every object and every life has to have its "when" of existence. The most logical definition of time is "From when to when." It implies existence. <u>Nothing can exist without its "when" (that is the moment or time) of its existence. And if this time is not naturally existing to which all mankind is born, as an objective entity, then serious philosophical issues arise.</u> But the passage of time is not the passage of that existence. It is rather additional, secondary, or complementary to the existence; something we use not something by which we move with physically in tandem. We remain where we are (you may stand or sit on one spot for hours); only the shadows moving over the existence constitute "time"---and only the shadows move, not the existence. To me, we've got the study of time all wrong. I am stressing this because it's so deeply ingrained.

Furthermore, how can anybody sit still as time passes by even though time does not move? The reason is that something is moving to give us time, but what is moving is not the time itself, being the successive hours or days. It is an instinctive mental illusion to think that time moves physically. It does advance but only arithmetically through succession. Time is a moment of contact with the external world (vision, tactile, etc.) If the contact or perception of the external world continues---say the length of daylight--- the time arises from counting something to give you how many 'somethings' (or numbers of it) you counted during that perception or contact with reality. Each unit of time is on its own but in procession with other units, though unrelated ('non-interacting units of time' is the logical name given to this process by Professor Whitehead). What are moving are the shadows of the

---

[16] That is the reason for calling Einstein a 'Philosopher-Scientist'.

astronomical bodies, since they constitute the time system. Thus you could sit still for hours, meaning counting the cycles or the regular motions used to reckon time. The motions are physical not temporal----but in psychology we count them as the rate of the passage of time. This is probably wrong, but we've got used to calling it time.

<u>Contrary to the concept of time passing through as the irreversible passage of existence, actually the passage of time is the time, for the passage of existence does not happen</u>. We're misled into thinking that the motions of shadows that constitute time means existence itself is being pushed by time or moving as time. This cannot be correct because existence is not one. It is multitudinous and move erratically---some are moving up, some are moving down, in reverse or even flying, but time moves steadily as programmed (in seconds or hours). These physical movements of existence (of Beings and things) are also purely physio/chemical and can only advance through chemistry. For existence to move in the way that time moves, all existence would have to advance by the same means---i.e. either in seconds or hours, days or months, as the case may be, yet that clearly does not happen and is quite impossible. Multitudinous Beings move in multitudinous ways.

However, it is true that even Bertrand Russell and Professor Eddington agreed with all the dire inferences about time coming from relativity. But then how did we get our time ticking away in the clock? This question will come up again and again until we find a logical definition of time that gives a cogent explanation of time in the clock. Also how objective or reliable is it? So there are genuine philosophical puzzles here. Nobody can pretend that they have been resolved, yet secular time which replaced divine time is also beyond dispute because it has been found in experiments that time can begin from anywhere. I will attempt my own solutions presently.

This is not a subject to be discussed lightly. It's taken me all this time to prepare this rejoinder. In fact, if religious people were half as clever as they pretend to be, they'd be more concerned with the relativity theory of time than the abuses from rude atheists.[17] Looking back, I recall another article by the great science writer, Dr John Gribbin, entitled "Pay Attention Albert Einstein!" (New Scientist, 2nd Jan. 1993, p28). These two articles have put in print so many controversial and contradictory assertions about time or post relativity time that I am lost for words. Yet once something is published it's almost impossible to refute it, otherwise Idealism would not still have millions of followers in the churches.

Dr Gribbin in particular asserts that the Minkowski formula helps the understanding of the special theory of relativity. He said: "Minkowski's geometrical description undoubtedly improved the clarity of the special theory and is still regarded as the best way to understand it."[18] He's not wrong; everybody thought so at the time, yet Minkowski came in years after the world acclaim of the special theory of relativity where space and time, in the words of Professor Bernstein, "were made separate and played distinct roles".[19] We are now told that Einstein did not even understand four-dimensional space or 4-D geometry, and I believe it---see below—for he had his own definition of space-time to begin with. There is no doubt that Einstein was very, very, very, clever! It's unlikely we'll ever see another like him. Not in the age of the Internet anyway, when the

---

[17] Personally, although I am probably the most irreligious person that ever lived, bar Bertrand Russell, I still consider it unwise to insult religious people---they cannot live without their religion. And in law they're entitled to believe what they like so long as they do no harm to anybody. Although some of them in the primitive societies do harm children, the problem is proof. Interference with public policy is another issue altogether; yet as strangers on a strange planet eking out a living as best we can till death, I wll always counsel mediation, reconciliation and amicable settlement of disputes, or 'live and let's live'.

[18] Ibid, p28.

[19] Albert Einstein and the Frontiers of Physics, Oxford and New York, 1996, p110.

half-wits are claiming that robots are going to take over the control of human life, assuming (wrongly) that any robots could do that without microchips of them being planted in every man's belly. And maybe they could do that to their own bellies, but I'm going to insure mine against that.

With Albert Einstein, it's different. We're still learning to understand relativity properly as the essay "Commentary on The Theory of Relativity" in The World of Mathematics makes clear.[20] First and foremost, time is not running through nature, as Professor Sir Arthur Eddington has categorically confirmed. Also there is no longer a universal time (Russell)[21], and time doesn't seem to move at all. And if that is so then time is our own creation/construction, Russell again)[22], and must necessarily be discrete, for Lorentz has proved by experiments that it can begin from anywhere. Let us clear these major points up at once.

ASTRONOMY AND TIME

The concept of time in experience may not be astronomical, yet it is caused by the motions of astronomical bodies. Take the sundials for example; they do not move but they can tell the time as

---

[20] Published by Tempus Books of Microsoft Press, 1988, Vol. 2, p1083: In this essay, the author makes it plain that we do not as yet understand relativity properly, so much for the British halfwit cosmologists who claim to be as clever as Einstein by merely speculating about the consequences of black holes---and adding insult to injury by saying Minkowski helps us to understand relativity. Actually he distorts the theory. All those who propound theories based on 4-D geometry are distorting relativity because it does not exist simply because time cannot be represented with 'i' in mathematics, let alone reality. So let us try to forget about the Transformation of Coordinates as it could end in fantasy. The Minkowski ict equation cannot represent true time, because time is not imaginary. That's the end of the matter with Minkowski, Time is difficult to define but it is by no means imaginary. Besides when you find a cogent logical definition you realise that it is required in all human activities, from going to the farm, work, travelling, schools in sports and even sleep, scarcely something that can be considered imaginary.

[21] From his ABC of Relativity---see below.

[22] From Mysticism and Logic as cited below.

moving unit by unit---they are units of our own making, which is another evidence that time is 'a construction'. Otherwise what is the time they are telling, or giving us? Where does it come from if there is no longer a universal time? This was the greatest question about time or post relativity time asked by Lord Bertrand Russell---the world's most recent greatest philosopher. He asked in his book ABC of Relativity Ch 4, (but of which most writers seem unaware), "If cosmic time is abandoned, what then is measured by the clock?"[23] The logical or scientific answer is nothing, *absolutely nothing*. Thus some writers claimed that time cannot exist under the special relativity metric---but we do have time! And we do live in a special relativity world or frame.

Since this is the most difficult subject in our intellectual life, let me stress the bone of contention again---and again---so that I cannot be misunderstood: what we want to know is the acceptable logical definition of time or post relativity time, if time is not fixed, divine, absolute nor generally covering or running through the universe and the same everywhere so that a second here is a second everywhere else. There! And we do know that this is what relativity time means---namely that it is not fixed or absolute, divine or general and the same everywhere. It varies from place to place according to the Lorentz discovery of local time, also known as t_1. I can now see why Einstein originally wanted to call Relativity 'The Theory of Invariance'.

As stated above in the notes, Einstein was not called Philosopher/Scientist for nothing. Indeed Bertrand Russell

---

[23] The original query is this: "If cosmic time is abandoned, what is really measured by a clock?"But I regard it as one of those clever sayings that can be stated in many variant forms with the meaning intact, and I enjoy teasing those half-wit scientists and mathematicians who fail to study the philosophy of science in depth. It's a serious issue, and no laughing matter. As a scientist you must start, as a universal duty, by asking why anything is what it is if not created by god. Finding the answer is your scientific research, though the question is philosophical; so anybody who disparages philosophy is lost. But to know philosophy you must start with logic.

considered his theory of space-time as (perhaps) his most important revolutionary discovery---and God knows there were many others. To me it is more important than his explanation of the cause of gravity. I don't care much for interstellar gravity; the stars can go on swallowing each other; it has nothing to do with us, but time has. (It is worthy of note that his concept of Space-Time was different from the Minkowski theory---for space was merged with time in the special theory of relativity but by the 3+1 formula: that is, physically not mathematically on the basis of imaginary time.)

Astronomy is what concerns us, not cosmology. That is an interesting intellectual game mathematicians like to play to entertain themselves when bored or lonely. Even if it can affect us there is nothing anybody can do about it. It's often forgotten how insignificant man is. In the wider cosmos we don't count at all. Even our sun, as huge as it seems to us, is really only a minute infinitesimal dot in the Milky Way and yet there are billions of such Milky Ways all over. This is the universe of complex events occurring in all sorts of bodies, some of them so huge that millions of our petty sun 'can find room in them'. As I write it is reporter in the papers that astronomers have discovered a black hole about 12 billion times more massive than our 'huge' sun.[24] So what is the status of this little animal called man? Philosophy is interesting and should be taught in High Schools.

WHY DURATION IS THE ENGINE OF TIME---THE CAUSE OF THE SENSE OF TIME

All the elements that cobble together to give us the sense of time passing are connected; but this is just like asserting that all things in nature are connected, and also all things and events are caused. So back to the sundials; it was necessary to clear up some related points so that it can be clearly understood. The sundials tell

---

[24] The Times of London, 26th Feb. 2015, p20.

the time by means of shadows. But that is all we can ever know of time and never the true nature of it; the time is driven by something (the real cause of time) the nature of which we can never discover absolutely clearly. The time the sundial is telling or showing is obtained from the duration of the earth's orbit of the sun, or the act of dividing the earth's orbit of the sun. 'Duration' means "during the time it is there" or through historical records (the same thing as 'experience'.) Let us use the simple word 'duration'. Since it means during the period of some event, it is always used with reference to something or some events. The orbit of the sun is one such events in human experience logically suitable for marking time; and we divide it into units of time (by space traversed) for cultural use---from seconds to minutes, hours and days, etc., and the sundial shows how these units are passing, and we are forced to obey them because as they pass they bring darkness and the rest of it that could spell doom. Hence we read the passing shadows as units of time and act strictly according to them. They are, in fact, units of space. For the more we probe time the more it becomes mysterious; to make sense we have got to stop at a point that is logically definable and say we simply do not know what the rest is. That is the best we can do through logical thought or science by courtesy of relativity---though presently ignored due to the influence of man's religious aspirations.

Thus a sundial is a clock; it tells or reflects what the time is. *Yet a clock which tells the time or shows what the time is solely by means of shadows means it is recording the passage of shadows as the passage of time.* So the shadows are giving the time—and constitute what we call time in essence. Since there is nothing else to call time, the shadows are, in fact, the time, or all we can ever know of time, and they are passing **so the passage of the shadows is time or the passage of the tor what we know as time.** without any need for theories to account for how time passes through nature, or before our eyes. This is the proof that all we can ever know of time is how it is passing by.

An additional point is that the shadows are caused by something close by, the closest causal agent may be called the 'primary' cause. Let us call the cause 'duration' for that is what gives us the sense of 'a period of waiting' that can be divided into shorter and longer periods. For instance, daylight is caused by the period 'during which' the earth is moving across the sun; if the movement is slow the daylight would be longer than the normal twelve hours; if faster, the day would be shorter and so forth. The duration of the earth's journey across the sun causes the time.[25] Thus there is no argument against calling duration the power that causes what we experience as 'a period of waiting' or time. When we are lucky we can identify the primary duration (that is, the nearest.) Otherwise it may be hidden: or seen to be caused by obstruction, motion, inertia, ebb and flow. Etc---and ignorance; for it is ignorance to say we have days of the week: there is only one day. The rest are human concepts for cultural convenience. I must mention here that cultural facts and natural facts are two different things; one is for physics and philosophers, the other is for human convenience.[26]

Thus we can imagine that there are other causes behind the primary duration, and on and on to infinity (Primary, Secondary, etc.) They do not concern us; what matters is that the sundial is a clock which tells what the time is by means of shadows; so when we read the sundial as saying it is ten O'clock going to eleven O'clock it means the shadows are moving and that to us is 'the

---

[25] As hinted above, duration, like the year, can never be defined except in reference to something else. Although unheralded duration is the greatest conundrum in the universe. To my mind it is existence but consists of layers upon layers to infinity.

[26] I agree that this is not a theory academics can respect because it is far ahead of anything they know; but at least they should reply to people's letters, since not all of us are humbugs.

movement' of time. So the movement of the shadows used to tell the time is what we mistake to be the movement of time---and time itself, whatever it is, does not move. However, since we know from QED that everything is caused, we can assume that the shadows are driven by primary, secondary and other duration, also caused by other forces to infinity. If cosmologists want a logical method for defining time without mythologies, here is it! Duration comes in because it is what we can divide and mechanise so as to be able to tell that one hour is longer than one minute.

Let me repeat for emphasis that the shadows that sundials record as the passage of time do move, of course, but not the sundials. So it's astronomical motions and shadows that we use to reckon time. Time itself does not move, as it's a mere psychological concept---i.e. counting the orbits of the sun as years is a matter for psychology not physics. We don't know what time is because the orbits of the sun are physical not temporal. We count the orbits to indicate the rate of the passage of time and never the real thing; so the passage of time is all we can ever know.[27] Logically it's impossible to discover how long the year is in duration. Nobody can define the year in logic without using any of its fractions; but then the fractions too have to be logically defined. It's therefore impossible to define time; and since the time units are created by ourselves out of our perceptions, Whitehead was right to consider time as 'instantaneous spread of the apparent world', *since the units of time we use or need are created with points out of the moments of time perceived, as already mentioned*.[28] So the best we

---

[27] One orbit, for example, means a toddler should at least be able to stand on his own, if not able to walk; from such instances, we work out what orbits of the sun mean to us in terms of time: one orbit is so-and-so, two is that, and all the rest of it---it does not give us anything else except that we call it time. In fact it is the rate of the passage of time only. So there is no need for theories to account for the passage of time, as I keep repeating and have to repeat to dislodge that aspect of our minds conditioned by the primitive notions of time.

[28] A mathematical equation for this is provided below.

can do by way of definition is to say time is the mechanics of shadows moving over existence, just as it happens over sundials to tell the time: you stay where you are and the shadows tells you that you've been on that spot for so-and-so cycles, hours or even days or years, like houses remaining where they are for ages. Ancient man was wiser than we think.

We don't know what time is (and it seems nobody can ever find out) but I suspect that, as a period of waiting, it has something to do with physical and organic chemistry, the natural processing in nature that causes us to wait (for some time in any activity.) The waiting period is what we call 'time'. For the waiting period is divided by the number of passing cycles to reduce it (however brief or long it may be) to units of time, and since these units are fractions of the year obtained with points, they can be called 'cycles', scarcely different from tapping the figure. But I hope the reader will also realise that I am repeating certain points for emphasis as the mind is otherwise so steeped in ancient myths about time that people would not understand what I saying set against what they call 'time', consisting of past, present and future. After all even serious academics say Einstein was wrong about time because he dismissed past, present and future as 'stubborn illusions'. Man is basically silly because of his emotional problems; for most of the puzzles in life may be soluble, yet men are emotionally prevented even from considering some of the solutions because they do not like the look of them; and unfortunately that's when religion comes in. When man is defeated by natural forces he calls on his God as 'The Father' for salvation; but once religion comes in death to the heretic is not far away, hence man's life is bedevilled by warfare.

WHY SECULAR TIME IS NECESSARILY DISCRETE

In sundials the motions of the shadows mislead us into thinking that it is the time that is moving. This is an excusable misconception because time goes from unit to unit, hour to hour, and year to year. Since the time units are fractions of the year and the year too is determinate, the time is bound to be discrete overall. The mathematicians were right all along, except that they didn't know how right they were and thought they're copying it from God---and all because of their feelings about the existence of God! That is the kind of human emotional burden I was referring to a moment ago. Many scholars are so consumed by their religious beliefs that even though they accept time as secular (after the Einsteinian revolution), but still regard time as running through the cosmos from the past to the present and infinite future (in phrases like 'the dawn of time' or 'end of time'), so time travel might be feasible. If they understand the Minkowski formula they additionally conclude that "curved space-time" makes time travel a certainty. Yet discrete time means the year is real. And if it is real then it must be discrete, for discrete time means from one point to another---and we always have to repeat the year to make our time continuous. Discrete time cannot move; discrete time cannot march forwards or backwards. It consists of moments of contact---long or short. Some of these are so long that we count cycles to know how long they last as they recur over and over again. In short, Professor Whitehead got the philosophy right. From Lorentz to Einstein, Russell to Whitehead, and the quantum theory to QED we can be fairly certain that we know what time is but very difficult to demonstrate it plainly and a lot of scientific imagination or deduction is required. Not just thinking, but thinking scientifically.

Again, in reality, only the 'discrete' cycles we use to reckon time do move; time itself does not move, so it cannot travel all through the cosmos like a stream or a thread. We don't even know what it is. We count the cycles or regular motions as the rate of the passage of time, the real nature of which is unknown. But those who argue that there is no time have a point, except that they do not make the point clear, or define it logically. For it appears that the movements we call "time's movements" are caused by the shadows of astronomical bodies, exactly as demonstrated by the sundials---that is why the passage of time is the passage of these shadows. Or the passage of the shadows is felt by us as the passage of time, since they bring the days continuously and the passage of the days has long been known as constituting the passage of time. In reality, there are not days in nature at all. There is only one day; the same shadows create the illusion of the days succeeding one another. And the hours of the day (or specific units of time) are caused by the slow movements of these shadows across the sky. All we do is subdivide them with points and give them names for cultural purposes---from seconds to hours, and the days too from Sunday to the next Sunday, yet as observed above, what is cultural is not necessarily present as such in nature. What applies to the days, applies with equal validity to the years and all the time in the universe, as part of the logic of time in the universe; every existing sentient 'Beings' will have to create a time system similar to the one described. Otherwise time does not exist; it is 'constructed' by sentient Beings. The factors and parameters required for constructing time sequences may be everywhere, but it takes intelligence to do the construction.

The best logical definition of time in science is "a period of waiting"[29]; but something causes the period of waiting otherwise there is no time as a separate entity. What causes 'the period of waiting' is time in the eyes of the person who experiences it. It is also time to the person who causes it, if it is caused by a sentient being; the need for periodicity is a natural law (and probably originates from the brain), and, to me, it is time; so it is not correct to assert that time does not exist---we couldn't even live without it, nor could we define existence without it since to be is to be in time as explained below.

---

[29] I borrowed this phrase from Professor Richard Feynman.

We count the mere shadows or cycles of physical movements of time for the real thing: we say we are aged ten years when we orbit the sun ten times---but the orbits are mere physical events! By this theory, indirectly, even the problem of the passage of time is solved----we only know how it is passing and never what it is; and our knowledge of the passage consists of our counting of the cycles, e.g. as 'years' pared down to the seconds and atomic units. The passage is the time or what we know as time and, I repeat, there is no need to invent complicated theories to account for how it passes by. This will not stop religious people anxious to come back after life but intellectually they no longer command respect.

Altogether, it's astonishing how Einstein was able to solve so many of our dire problems in so short a time. I believe it's because he's strictly a logical thinker (as Russell repeated many times in his many books), and through logic and the dynamic mechanics of atoms and mathematics, all things are connected; therefore science is indispensable in all human affairs, not just one way of looking at the world but the only reliable way in so far as all things consist of atoms. Even this last sentence is inferred from Einstein's ideas. So long as there is physical elements in anything, there is bound to be some aspects of science in everything.

As noted above, there is no longer a universal time covering the whole cosmos, which to me is the most serious philosophy of nature since Copernicus. This is how Russell put the notion: "There is no longer a universal time which can be applied without ambiguity to any part of the universe; there are only the various 'proper' times of the various bodies in the universe". For Einstein, (as Abraham Pais put it in his book "Subtle is The Lord..."): "... there are as many times as there are inertial frames. That is the gist of the June paper's kinematic sections, which rank among the highest achievements of science..." He went on to plead that it should be taught in schools, and I agree with him.[30]

---

[30] Subtle is the Lord, by Abraham Pais, Oxford, 1982, p141

Next, Professor Sir Arthur Eddington. He wrote: "Prior to Einstein's researches no doubt was entertained that there existed a 'true even-flowing time' which was unique and universal...Those who still insist on the existence of a unique 'true time' generally rely on the possibility that the resources of experiment are not yet exhausted and that some day a discriminating test may be found. But the off-chance that a future generation may discover a significance in our utterances is scarcely an excuse for making meaningless noises".[31]

Given these categorical notions of time as a secular entity, the definition of time has become very philosophical and difficult, making it as close, essential and inseparable from life as water. But how does it (or did it) come about if not from the Heavens? The consensus in science is that it just is. Maybe, but even so how do we define it?

Our own Lord Bertrand Russell was the world's greatest philosopher living--- Logician, Mathematicians of genius, writer of genius, and as clever as, or more so than, Aristotle. For Aristotle had demerits, Russell had none, just pure brilliance; and as for science, he founded the philosophy of science with the book "Our Knowledge of The External World". He it was who defined post relativity (or secular) time thus: "It seems that the all-embracing time is a construction, like the all-embracing space. Physics itself has become conscious of this fact through the discussions connected with relativity."[32] Yet it failed to give us a clear philosophic idea of what time is and how it began as a secular entity----people just ignored it and continued to regard time either as just is, or divine, with most of them making fortunes upon fortunes for regarding it as linked to the Minkowski space to make it possible to travel by curved space---hence the popularity of time

---

[31] The mathematical Theory of Relativity, Cambridge, 1930, Ch.1.

[32] Mysticism & Logic, George Allen & Unwin, 1917, Ch viii (x).

travel or the notion that time and space constitute one entity and the mystery of time deepened---otherwise Time Vs Space as a serious topic for publication would not be entertained in science.

WHAT A MOMENT MEANS

A moment is any sort of contact with nature, meaning all acts of perceiving (visual, tactile, etc.), so long as it is determinate and has to be repeated to continue---no matter how long it lasts. And that is the interesting point, because it can be divided and still be a moment or part of a moment.

At the time of Russell and Einstein Professor A.N Whitehead too was alive, and he was some brain. He defined time originating on this or any planet as post relativity time, implying that divine time could not exist. In The Principle of Relativity Professor Whitehead wrote: "...a moment of time is to be identified with an instantaneous spread of the apparent world"--- in other words, a moment of perception, vision, existence or 'Being', and went on, "...A time-system is a sequence of non-interacting moments [however that moment is defined]".[33] He was not a very lucid writer, but I suppose this is what he meant: every unit of time is a moment in life. Of course some of these moments are very long, like the year, or day, but they are moments in the sense that they are determinate units or periods that have got to be repeated to continue. This is an attempt to find a definition for time that suited its new status as a secular entity, created or constructed (in the words of Russell) by ourselves----the year for instance, pared down to the seconds or even the atomic units of time which have always to be based on the second to make sense. Otherwise there is only one day and only one year; so any time system based on them is bound to be discrete.

---

[33] The Principle of Relativity, Cambridge, 1922.

This is where the problems began because discrete time cannot run all through the cosmos in the form of a thread or stream; discrete time cannot bend, move or march; discrete time will not make time travel possible; discrete time does not cause the story of history, which is rather seen as the march of events not of time for only events can move forward, so that we carry the past with us always to the present and the future. Thus Einstein was right: past, present and future are mere linguistic illusions: you live with your past and will carry the same to your future---examples are everywhere, your bank balance for instance! You do not have to visit the past to access your bank balance; and the balance tomorrow will inevitably be what you have today----the bank manager will not pile pounds into your account for no good business reasons.

I honestly cannot imagine how anybody of whatever status, intellectually, academically, politically or religiously, could contradict the reasons for secular time sketched above to claim (by revelation or whatever) that time is other than the post relativity concept of it. Yet what keep appearing in books and magazines are still concepts of time in the old format: that it started from Time Zero, it just is; it increases, runs faster, slower and so on---yet discrete time cannot do any of these things; and since our time is based on the year every year, it is none other than discrete. I am often confused when scientists refer to something called "The Dawn of Time" unbroken to this day. And they are fond of demonstrating this by arithmetic as they count the years. Shouldn't it be "The Dawn of Existence"? According to all the authorities cited about time or post relativity time in this book, there appears to be no "unbroken streams" of time running all through the cosmos from the Dawn of Time or Time Zero---isn't that what relativity time means and isn't the contrary idea exactly what Professor Eddington called "meaningless noises"? Otherwise there would have to be different streams for the various units of time as we couldn't have one stream of time running as seconds, minutes, hours, days, months and years---quite impossible to programme into a clock. Even the traditional definition of time acknowledges this. It speaks of "The passage of existence". In a multiple stream time it would be "*The passages of existences*"---plainly an illogical notion, as it would be a world of intermingling streams of periodicities all passing away at the same time with different velocities or momentum.

I must stress again that it's not right to say time does not exist, but nobody can define it logically other than as 'a construction' out of the features of the earth and other astronomical bodies. Sir Arthur Eddington was a very clever mathematician and scientist of genius, the founder of Astrophysics, and he says any such ideas about time after Einstein are fatuous---"meaningless noises", as he put it. That should keep the 'Doubting Thomases' quiet, at least for now. To help them along, I give below a brief sketch of the new theory---already I have shown that it solves the passage of time in a flash.

Now, currently professor Palle Yourgrau of the United States is making a name for himself as the champion of Time Travel, because he has written a new book called A World Without Time (Penguin, 2007), in which he claims that Time Travel is 'a scientific possibility', (which is evidence that the old theory of universal time running all through the cosmos is still prevalent in a great deal of the academic world). Time as a moment of existence (no matter how it is perceived) whose succession creates the illusion of continuous time can be incorporated into science and is backed by experimental results, Bertrand Russell, Professor Whitehead, Einstein, Professor Eddington and Gottfried Leibniz. Eddington even said (probably angrily) that opponents of this theory are making "meaningless noises". However, time "as just is" (not given any definition), and which began at a date chosen by the writer (like Archbishop Ussher) is the bedrock of mysticism and has no place in science---yet scientists seem unaware of this. For fifty years my manuscripts are never even read before rejecting them; they'd ask for a summary and reject the work they haven't even seen. The reason, I fear, is that everybody wants to believe that time travel and life after death may be feasible with the mystical theory of time, so that life will go round and round through the cosmos forever. I am afraid we are doomed, not safe even in the hands of our own scientists! Pythagoras is responsible.

Nevertheless, crucially, Professor Yourgrau has provided evidence that Einstein did not, in fact, even try to understand the Minkowski theory of four-dimensional continuum, or, in plain language, the theory that space and time constitute one entity: that Minkowski has linked (or equated) space to time.[34] He wrote, and I quote: "Every boy in the streets of Gottingen understands more about four-dimensional geometry than Einstein. Yet, in spite of that, Einstein did the work and not the mathematicians."[35] He himself quoted this gem from David Hilbert, therefore we can be certain it is true. Yet we know that Einstein used the Minkowski formula in his general relativity. The presumption then must be that he did so just to placate his mathematical critics who were calling for him to be hanged by the nearest lamp post.

Technically, it is quite impossible to equate space to time by means of mathematics unless one relies on 'i'; but then the theory is vitiated because time is not imaginary and 'i' can only be used to represent imaginary quantities.

The creation of space-time, being the merging of space and time (still as separate entities), was achieved in special relativity, as Russell has observed, by the use of the 3+1 formula, but

---

[34] One has to admit that if this is true then the world has changed out of recognition, and time travel would be possible. The irony is that it is not correct yet the world has changed through Einstein's own theories of time, namely that time does not run through all nature and the same everywhere, and that our own time was constructed by man.

[35]Palle Yourgrau, A World Without Time, Penguin, 2007, p6.

mathematicians were spitting blood because time had been made secular at the same time. They said it means man (as emotional, biased, partial and fraudulent as he can be!) creates his own time and then adds it to phenomena and call it objective reality. That's not an acceptable concept to represent true reality, they claimed. Yet 4-D geometry does not and cannot exist, so the theory of curved space-time is flawed, and time travel via Minkowski is not feasible. The history of the Minkowski effort is interesting even before we come to the story that Einstein did not understand it, for saying that in reference to any subject in physics or even science generally, means the great man thought whatever it is was nonsense. But Professor Eddington called it arbitrary and fictitious (though useful for the study of phenomena in his Mathematical Theory of Relativity). Russell said it was compounded for the convenience of mathematicians (The Analysis of Matter). And one reference work (at least) called the theory artificial (The Routledge Concise Encyclopedia of Philosophy). It is true, of course, that Mathematicians adore the Minkowski theory because it enables them to dispense with the 3+1 formula, which they regard as less than objective for science.

In my judgement, I suspect Einstein did not even bother to understand 4-D geometry. I can imagine what was going on in his mind. The mathematicians had to be placated and come to support his theories so that he could get on with his work. For that purpose he praised Minkowski and pretended to adopt his formula. The whole world was misled into thinking that because Einstein praised Minkowski he had a hand in the theory of relativity. He tried to contribute to it but failed because he had to base his theory on imaginary time coordinates. So, in fact, the Minkowski formula was irrelevant---there is no way it could have vitiated the new theory of gravity, and I would bet my last penny that Einstein knew that. The definition of time was a different issue. He said (and I quote from Abraham Pais's Subtle is The Lord...): "All that was needed was the insight that an auxiliary quantity introduced by H.A Lorentz, and denoted by him as local time can be defined as 'time', pure and simply". The many times this is quoted reflects the stubbornness of mathematicians!

Finally, as already mentioned, in the absence of a universal time, Russell demanded to know what is measured by the clock (ABC of Relativity, Ch.4&5). In fact, there is nothing. So what is time? Nobody knows what time is, but by using the Lorentz concept that time can begin from anywhere, with Einstein's support that there are as many times as there are inertial frames, we can logically infer that time is constructed by man and that it is purely psychological and cannot exist outside the human mind. Mathematics is no help here. Generally mathematics is indispensable in physical matters not in matters of the mind.

Yet, in the end, time is basically counting cycles; therefore it's mostly psychological. Ten orbits of the sun are ten years. That's true, and in the absence of a universal time nobody can define time besides the mental notation of the orbits as time (the years of course are the years, the measure of our own ages); but crucially somebody must be there to set the points for the yearly cycle or there will be no years and seconds derived as fractions of the year; so it is also true that we create the years or time, as Russell has pointed out. The theory of the new concept of time is that any cyclical or regular motions divided by points will provide periodic intervals or time units for cultural use as the logic of time in the universe; that is how we get time to put in the clock. We can simplify this in mathematics thus: $RM.P = TS + E$, meaning any "Regular" "Motions" "Divided" with "Points" provide "Time Sequences" as the logic of time in the universe. Beyond that there is nothing for the clock to measure as time units. The 'E' represents "Existence": Regular Motions Divided with Points provide Time Sequences for defining Existence---defining, justifying, legalising etc. The logical reason is that (culturally) everything in existence has to have its "when" (or time) of existence in the universe. For while there is no universal time, there is nevertheless a universal law for time and existence as the definition of life: every 'Being' has to exist or be covered by time so that it can be defined as existing at so-and-so a time or it never existed. Every existence can only be quoted in time, otherwise how can it be cited? This being so, the above equation becomes the equation by which all life can be defined or justified either in law, philosophy or science. That's how important time is, and we can say that it's deduced from Einstein's ideas.

It must, however, be noted that time is backed by, or based on, duration, the real unknown mystery of time, which enables us to tell that one hour is longer than one minute. The logic (or the law of time in the universe) arises from the fact that duration has got to be divided to obtain the culturally indispensable units of time, since all time is known and used only in units. The word time means nothing in culture without quantification---meaning 'the quantity of time' involved. Let's suppose that you are asked to sit down in one place for one hour as punishment; other persons are to do so for five minutes each. As you sit there people come and go rapidly every five minutes. You're ignorant of time units and their various durations, so you begin to wonder what is the difference between five minutes and one hour. In other words, how do we differentiate the lengths of duration between the different time units? In all the universe there must be a law or method for doing that, and, 'speculatively', I believe it is this: we count cycles based on the points used to divide the earth's orbit of the sun, and that is tantamount to counting shorter cycles with points out of the total duration or length of the orbit of the sun. We have no other methods for acquiring units of time; and this will be true of any determinate cycle used for reckoning time in any part of the universe. For the discrete nature of time is caused by the determinate cycle upon which it is based---year after year after year. There is only one year; we repeat it to get all the years we speak of. And since the year will end and restart we have to divide it into an exact number of units to coincide with the end of one year; this can only result in a time system that is essentially discrete with all the momentous implications of discrete time---something originating from and ending with man, which shows the nullity of all the prognostications of all the religions for a start! Time is not an easy matter; it controls everything. The irony is that it is 'constructed' by man, as Russell noted about a hundred years ago---we actually do create our time, though from factors naturally existing in the universe.

Duration (that is, during the period a thing is there or is encountered), of course, is natural and existing all over the universe and (under QED) it must be 'caused'; but to exist (or live in an inertial frame and have culture) you have got to find a mechanism for quantifying duration into time sequences, dividing it into manageable units. So duration caused by many factors (inertia, motion, atomic and nuclear processes, ebb and flow, force, obstruction etc.) must be existing all over the universe, but it takes the human mind to invent time sequences out of it. Luckily for scholars, for once it is plainly not a chicken-and-egg question since the factors, conditions and parameters for 'constructing' time sequences are all present throughout the cosmos. But since the construction requires points we had to come down from the trees to learn to do so. The year, for instance, is determined from one point to another: from midnight 31st December to the next midnight 31st December and then start another year on and on forever. Even one second before midnight is this year, one second after and it's the New Year---thus it must be recognised that the second is part of the yearly cycle. Intelligence or sentience is required in time's construction, plus a theory of numbers, the ability to count and arithmetic. Time does not exist outside the human mind, which, of course, means another line of inquiry must now begin!

Thus time, as important as it is, is an artificial contraption based on existence for the justification and regulation of that existence and all activities. Let us say we could use the hand round and round without muscular strain. Ten cycles means it is time to go to school. At school, a hundred cycles means it is time to go and play, another hundred cycles and it is time to go back home---something like this suggestion (as a mechanism in a clock) can be seen as a device for reckoning time to regulate activities. For time is the same thing as existence or Being, and we use points to regulate the existence's activities. Otherwise time and existence are inseparable; there always will be time, the problem is how to express it (mechanise and control it proportionally) in a general form for the regulation of all activities. That does not require complicated or counterintuitive mathematics----otherwise how can people understand it? It is naturally there and inescapable, we only have to regulate it. Take the daylight for example. All we need is something like the sundial, where the positions of the shadows of astronomical bodies (or whatever) enables people to know how to go about their various activities. This should not be interpreted as the motions of time; that it moves from morning, to mid-day and to the evening and nightfall---not at all. Only the shadows do so move, but it has traditionally been assumed that the motions are those of time itself.

THE IMPORTANCE OF PHILOSOPHY

One last relevant point is the nature of philosophical reasoning. Sadly, only philosophers can understand philosophers. We have to try and solve this problem. A philosopher has got to be interpreted; yet they do so as philosophers and in the language and style aimed at philosophers, so that the public is largely excluded. For instance, when Leibniz said that time is succession (Gottfried Leibniz 'letters'); and Russell also said it is relation between points (The Analysis of Matter, p376), both were providing solutions to the passage of time without the mythical arrows, but we couldn't understand them. Yet, as explained above, it is true that time is a succession of moments, or a succession of the relations between points, in the manner we get the years. Between points we get 'units', intervals or moments. The moments of Whitehead mentioned above can only be 'relations between points'---the yearly cycle for instance. Thus both Russell and Leibniz were right: the passage of time occurs through the succession or procession of time units, intervals or moments, exactly like the days and years. The day is only one and the year is only one; both are replicated (with different cultural names) to make time continuous, meaning they are time units, intervals or moments in succession, from the smallest unit, the second, to the largest---the year---they succeed one another to continue and the arrows of time are no longer needed. When time is defined like this, there can be no time travel, no divine time, no universal time and no physical passage of time---it is purely mental or arithmetic (the counting of cycles as time units), and what does not physically exist cannot be physically passing by. Above all we get the mechanics of this time to programme into the clock.

My definition of time as something we invent or construct to measure duration—the real mystery of time-- can be inferred from the following scenario: let's say you are blind and you touch something (that is Professor Whitehead's 'moment' of perception or contact with the world). When asked 'for how long?' and simply because you are blind and cannot read the clock, you 'construct' your own time (Russell), and say, "I tapped my figure ten consecutive times". That's time without mythologies, arrows or divinity. Another mundane illustration is a search light, a powerful one shown on a particular spot, say, from a-search-and-rescue aircraft as happens all too frequently. If one wants to know 'for how long', tapping the finger will do---that also gives the length of time, without knowing what time is. That is the important point. We can only use some kind of repetitive motions to indicate the passage of time without knowing what the nature of time really is; we count the motions and say, wrongly, that that is time going or that it is the motion of time; in fact it is the cycles going, the motions of the cycles. If we want to provide an objective method to be easily copied and applied by all human beings, we use the obvious one, namely orbit of the sun that affects us all equally and therefore unlikely to be missed.

All time is a period of waiting, but all such periods are also mysteriously caused by primary or secondary duration we don't see; they are what power the causes of perceptible duration. We say Day&Night is time going, but there is only one 'constant' day from the sun. The sun is on all the time. Also gestation is a period of waiting; it is caused by duration we cannot see---the chemical period required for foetal growth. The causes of duration may not be apparent, because most of it is astronomical or hidden chemical processes. There is even ultimate duration---i.e. the nuclear process in the sun and beyond. Duration is caused. The clock is a technique for measuring the duration in specific units (hours and minutes) for cultural use.

For scientific thought, we have to define time in one of two ways to accord with the mechanisation of time in a clock as we already have it:[36] either we say existence is time and use regular motions (because time requires points), to obtain our time units (the seconds and hours, etc.); or say the cycles and regular motions we use to tell 'how long' the existence is there (or has been there) is the time. But neither of these can be called the passage of existence due to the multitudinous nature of objects and their millions of

---

[36] This is to answer Bertrand Russell's question about how we get time in the clock in the absence of a universal time.

modes of passage through nature. Time passes through only one basic unit----in our case the seconds; but just fancy how many objects there are and how variously they pass through nature, mostly chemically. So we have to humbly abandon our love for theories and rely on pragmatism as primitive man did, and acknowledge that the cycles we use to reckon time are the true nature of time; their motions are the motions of time and they are what we mechanise in a clock. So we can continue counting the orbits of the sun as years without missing any aspect of time, except to make the study of time parts of astronomy and chemistry, for it is extremely important.

I know, of course, that many critics believe that time is just is and should be left alone—but even so, in what does it consist, or how should we define it logically? And we have to try, given its importance. Physics and chemistry need the clarification; it's not just a philosophical pastime to entertain logicians. The quantum, for instance, materialises after a period of time---whose time is that, and how do we define it? Is it our own time or natural time? If it's our own time (created or invented by man on this planet), then what is the status of the quantum of energy in the universe at large?[37] I don't think we can find that out using vague theories of time. For I have always believed that the theory of everything, even if necessary at all, cannot use the Minkowski notion of time in its composition (in any case, giving something in nature a name before looking for it seems queer to me). In other words, if the Minkowski space or universe is unreal (and 4-D geometry is a myth), then we have to be honest enough to concede that we simply cannot link any condition in nature as real, substantial and powerful as gravity to sheer myth, which brings to mind the Platonic simile of the cave

---

[37] Can some other 'Beings' in other parts of the cosmos using our mathematical proportions of our second as we do on earth discover the quantum and QED in their own parts of the cosmos? If not then we should not look forward to meeting aliens here---we could annihilate each other!

analogy with wider implications for all sorts of reasons, including dark matter and dark energy. Some people may even include the ghost world and the milieu of God wielding influence without being seen! Altogether man is very, very insignificant; the problem is the infinite searching power of the brain which forces us to be vain---so what are we, or where did the brain come from? The nature of the passage of time and the existence of the universe are not even half as mysterious as the probing powers of the brain. That's where scientist should concentrate their researches not the black holes and time travel which, even if true, have little to contribute to human welfare.

In conclusion, let me expand something I said in the notes above about philosophy and the academic scientists. Generally (when they're not mocking everybody else), they now treat philosophy as some kind of academic game anybody can play. So soon after Russell and Whitehead, this is very annoying and a disservice. For we all know that you have to have the necessary skills to play any game.

For instance, in his lavishly produced book The Universe in a Nutshell (Bantam, 2001, P.24), Stephen Hawking claims that he and Penrose "showed that general relativity predicted that time would come to an end inside a black hole..." This is an unnecessary and stupid academic guff---but even then so what? And is that a Nobel discovery even if true? We don't live there. Professor Eddington has already told the world nearly a hundred years ago that Einstein's researches have proved that time does not flow through nature and that to maintain the contrary view is making meaningless noises---The Mathematical Theory of Relativity, ch.1.1, as already quoted above. So what kind of time is Hawking and Penrose talking about if not the old religious notion of time? And how does the time begin and how will it march all the way to the black hole? Hence I can also predict that these so-called brilliant scientists in Britain (or is it the planet?) since Einstein can never define time logically in any language or mathematics. It's all the fault of Hermann Minkowski. And we now know that Einstein did not even bother to understand his theory. We have to treat relativity carefully because Einstein went very, very far than anyone has ever done or ever will be able to do on his own. Relativity and Einstein's ideas as a whole are not just physics; they're much more serious than that. Those who believe that philosophy is useless should realise that Einstein could not have done so much in physics without thinking philosophically about time, space and matter. For centuries religion was barring advance both physically and intellectually. So we have to accept that if Einstein was frustrated by such religious nonsense, he could not have done so much.

CHAPTER TWO

We live and can only live according to our time system.[38] This makes time second in importance only to life, no wonder the religions insist on their own interpretation of time. All through human history, and beginning with the Day & Night system (sleeping at night and working during the day), we have lived according to our notions of time. It's so oppressive that in many cases we simply have no choice. Since all calculations of time are based on the motions of the earth, and since we can only live by time, we live and can only live by the conditions of the planet, as a result we are compelled to live according to time; it is virtually unavoidable: conditions of the planet are determined by its movements, and these then impose a time system that is virtually unavoidable so long as we live on the planet.

For instance, the whole of the quantum theory is based on time, but on whose time, since there is no longer a universal or cosmic time? This is an ancient problem in philosophy, and we're

---

[38] If the brain were to reverse its natural functions and induce sleep rather in daytime, society would be organised to take account of that instead of going to sleep at night.

not even ready to face it yet. We think the quantum is the 'smallest bit of matter that can exist'. Yet it is time-dependent: so much energy emanates only after the lapse of so much time---I am trying to avoid the technicalities that put readers off.

The whole of our life is dominated by quantum mechanics as we learn from QED. But in the absence of a universal or cosmic time, whose or what time underlies all this?[39] How does it get mechanised in the clock? The answer is rational time, the logic of time, or logic and time, otherwise known as 'the quantity of time' or quantified time units, like the year pared down to the seconds.[40] It's the only way to acquire time. Before that insight we simply did not have a clue what time was. However the repetitive cycles we call time (say, the years) are merely physical events. They can only show how many cycles (years) have passed. If we translate that to 'how much time?' it means we can only know how time is passing and not what it really is.[41]

Now, if we ignore Archbishop Usher's infantile, religious time of Creation as we must, then there have been four intellectually

---

[39] It's not appropriate to call it duration and argue that duration can be inferred from the activities of anything, as I imagine some people might be inclined to do. Yes, of course it is duration but one translated into numerical units of time---but by what method? If it is our own methods then the quantum might not be a universal unit of energy, but peculiar to the earth's time system. This is only one of the many problems currently facing physics.

[40] See the meaning of these terms below. When I state that the year is pared down to the seconds, I include the atomic pulses or oscillations used to mark time since they have always to be related to the second to make sense: the radiation pulses of cesium 137 are used to measure the length of the seconds more precisely. That is what we call atomic time. It's neither atomic nor time. It's mere physical motions counted as the rate of the passage of time, and in any case it is based on the second which is a fraction of the earth year and therefore part of the yearly cycle.

[41] To my mind all time is chemistry, but that is another aspect of time we may never grasp, certainly not before a thousand years have passed. Our present view of time has taken thousands of years to condition our minds; it cannot be rubbed off within a few years. It's the old story of human life: tradition versus new knowledge, science and religion, philosophy and mysticism---life and death!

respectable time systems in the world overall. The first was just a process of people marking signs on the wall to show days and nights as they passed by or succeeded one another. It included the Sundials and many other primitive practices. (The concept of time 'passing by' is basic to all systems of time ever invented.) It ended with the Newtonian Absolute time, as our second system of time more intellectually respectable than any system before it. Then Lorentz discovered time dilation through physical experiments implying that absolute time did not exist; he called it 'Local Time', or t^1, and refused to believe that it was normal time.[42] Later he confessed that he might have been able to discover special relativity if he had paid due attention to his discovery---which Einstein was clever enough to pounce on to work out his theory of frames and special relativity, et al.

The fourth system of time was the Minkowski four-dimensional time, or 4-D geometry, which was purely mathematical. Everybody knows how cheekily he introduced his contribution in his Raum und Zeit (Space and Time) lecture in Cologne, 21 Sept. 1908: "The views of space and time which I wish to lay before you have sprung from the soil of experimental physics, and therein lies their strength. They are radical. Henceforth space by itself, and time by itself, are doomed to fade away into mere shadows, and only a kind of union of the two will preserve an independent reality." We're all taken in by Minkowski's confident proposal, including Einstein, or so it seemed. Foolishly, I once wrote a small book entitled The Mathematical Theory of Time based on the same Minkowski theory and then withdrew it immediately before philosophers of science had time to send the men in white coats after me. Not a single copy was sold because I

---

[42] The technical name of the Lorentz discovery is "The dilation of Time as a measure of moving clocks".

realised that it's rubbish or based on an illogical proposition. But at the time everybody thought Minkowski was a breath of fresh air.

According to the brilliant British science writer, Dr John Gribbin, "Minkowski's geometrical description undoubtedly improved the clarity of the special theory and is still regarded as the best way to understand it."[43] Well, if so then I am not surprised that some writers are saying even now that relativity is not properly understood. Set that against what Professor Bernstein wrote: "In the absence of gravity space and time are distinct entities. In the metric of special relativity they play distinctive roles..."[44] Time can be merged with space but in the 3+ 1 formula not by mathematics alone---for it all depends on how time is defined. Minkowski did not define it but obviously his theory implied that it is something in general existence. On the other hand, my suggestion is that, as Russell was clever enough to note about a hundred years ago, under relativity time is 'constructed' by us. The elements for doing this construction may be everywhere in the universe; but it takes human intelligence to do the construction.

Since there is no longer a universal time, there is bound to be a method for acquiring time, or what we use to guide our actions and which we know as time sequences. This method is what I am in the habit of referring to as 'the logic of time in the universe', in the sense that any 'Beings' in any part of the universe will have to invent a time system by the same method; and I imagine that the method or logic will include the following line of thought. Nobody can ever know the true nature of time. But events, action, effects of actions and sequences in all activities, are never continuous in so far as breathing is not continuous and will always require a

---

[43] Taken from "Pay attention, Albert Einstein!" By Dr John Gribbin, New Scientist, No. 49, 2nd Jan 1993, page 28.

[44] Professor Jeremy Bernstein, Albert Einstein and The Frontiers of Physics, opp. Cit., Oxford, page 110.

regulator, or the regulation of intervals; this will involve delays, motion, persistence and revolutions or any system that leads to 'a period of waiting, between activities', which is the same thing as time intervals---however they are caused. Once a mechanism is created to allow for this, time is created, and I assume that every Being in the universe will have a time system similar to that. It means using any cyclical motions and counting them as the units of time to guide any action that is not continuous like motion on a conveyor belt. This is the logic of time in the universe.

CHAPTER THREE

In a short digression to consider what Dr Gribbin has stated above, I would like to comment on relativity and the mathematicians' interpretation, regardless of what Einstein has said about the matter. I have received several dissertations about Minkowski and his equation of space to time. In one of these, the writer wrote the following in his attempt to explain the gist of the Minkowski formula: "Minkowski's space-time is his way of interpreting Einstein's Esynched system...Esynched clocks obey Lorentz's local time equation...in which t is the *general time*... and because the 'time' per consecutive 3d point is different than that of its fellows in the direction of motion, the 'time' in that system has a continuous range from infinite past to infinite future. That is the cause and meaning of Minkowski's 'four dimensional space-time continuum'". You'd need ten PhDs to understand this, and if you do, then you'd have to inquire from the writer what is measured by the clock as time for general use? For example, the writer mentioned 'Space-Time' yet he couldn't define it clearly. Here is how the great mathematician and logician, Bertrand Russell, defined or explained space-time for everybody to understand what it means: "SPACE TIME, as it appears in mathematical physics, is obviously an artefact, *i.e.* a structure in which materials found in the world are compounded in such a manner as to be convenient for the mathematician."[45] You won't need a single 'O-Level' to understand this, meaning that it is not true of the external world and so those mathematicians regarding it as such are wilfully distorting the theory of relativity and mathematical physics as a whole, 'wilfully' because they know it's logically untenable. A word of advice to inexperienced mathematicians: philosophers take all mathematics into consideration before deciding on the nature of the external world. So the bizarre notions dreamt up in mathematics are most unlikely to render existing theories valueless, but they can be consistent with them---improvements are welcome, but revolutionary ideas from mathematics are not common. That, precisely, is the problem between Einstein and those mathematicians yearning to prove him wrong or grab a piece of his fame. Einstein's ideas constitute physical intuitions or insights about the whole universe. You can apply mathematics to them; but you can never undermine them with mathematics alone---the physical constituents would have to change for that to become possible.

All those (philosophers, logicians and physicists) who think deeply about relativity will get a sharp pain in the mind that (defined like this, as our anonymous writer has done above), relativity will never be understood by those with professional, obligation to write for their fellow mathematicians' understanding of the theory, which is completely wrong and contrary to the philosophical interpretation of Einstein's ideas. Russell said space-time is artificial and compounded for the convenience of mathematicians. That's not an honest search for the truth, especially when the concept involved leads to a view of the world

---

[45] Bertrand Russell, The Analysis of Matter, Ch. XXXVI

that's the complete opposite of physical reality---time is not the same thing as space, because the equation to make it so was not successful, as Russell has confirmed. Therefore space-time, used in the sense that space is the same thing as time and vice versa, is a distortion of relativity. Einstein did not just write on physics. He reconstructed the whole of physical reality. A hundred years have gone and we still can't understand his ideas properly because writers are picking what they can understand. Thinking about the whole of physical reality is beyond ordinary writers no matter their status in the universities---it requires a team of great philosophers to interpret him properly. And where are they? We have had but only one great philosopher, Bertrand Russell, but he's dead. It's most unlikely we'll get another one because the world of learning has changed: nobody is encouraging such thinkers; even publisher complain they have no expertise in these matters, and rather prefer to heap millions on convicted murderers to tell their stories for commercial gain. In the past it's different. We had publishers with genuine love for knowledge; today there is only genuine love for money, even promoting tarnished sources of money with criminals and murderers. People regard Einstein as just another scientist. That's a mistake. In all history he is the only man to discover how to think logically about the world.

There are several specialists in science. But relativity is different because it is a logically deductive system of thought about the external world as sketched below. Of course the one theory will not be able to answer all of the countless queries about nature; no theory can do that; but for those professionally bound to work with it, I suppose, they can only understand the theory's contribution by thinking about it as a logically deductive system, beginning with the fact that *there is no longer a universal (or general) time due to the discovery of 'local time'*[46]*; that is the genesis of the theory of*

---

[46] For this local time notion is what revealed for the first time in all human history that

frames and it is upon the theory of frames special relativity is built; and upon special relativity that general relativity is deduced.[47] Einstein divided the universe into two: one is where life is feasible; the other is the metric of general relativity where life is not feasible, and he even went on to sketch the physics applicable to each half more or less perfectly. That is the calibre of the man whose ideas are handled so carelessly by mathematicians because they think they know best, if so why couldn't they discover relativity? Because of time, a great deal of humility is required in the study of relativity, or Einstein's ideas as a whole. Placed under philosophical scrutiny, the new concept of time, properly understood---without the Minkowski fiction---changes reality, human existence, philosophy and how we see ourselves completely, simply because time controls everything.

And there simply is no general time that can be manipulated with mathematics to "range from infinite past to infinite future". If mathematicians want to preach religion they should go to the churches. No wonder the Encyclopaedia Britannica (Macro) states that almost all cosmologists agree that space-time is infinite in its timelike directions[48]---what will happen to this time when the earth ceases to exist? What will the word 'infinite' mean? Are we to suppose that it will fly out of the window to the universe singing

---

time is neither absolute, fixed nor general, covering the whole universe and the same everywhere so that a second here is a second everywhere else. Everything in our intellectual life changed with this theory---yet it was not only a theory. He discovered it in physical experiments. It's the truth; and Einstein had the insight, the wisdom, the genius, to build his ideas on it. Nothing can be more scientific, and nothing can be more indisputable---the reason Minkowski must be accused of distortion. The fact that his mathematics is adored is even irritating.

[47] It is said a farmer went to Bertrand Russell one day and asked him to explain the theory of relativity in one sentence. This is something like that.

[48] Is it any wonder that the Minkowski formula is the principal inspiration for most of the theories about time travel in the infinite future or infinite past in which people could meet their grandparents even before they're married?

the praises of Hermann Minkowski? Or perhaps time will continue in people's heads as they travelled round and round the universe looking for a new home as Pythagoras supposed? I accept that life is harsh, sad and brutish as well as being short, as the poets have often told us, and death is simply unspeakable, especially when one thinks of those left behind who're vulnerable. Unfortunately that's the curse of life, and yet we never ask to come to life; it's always someone's decision to bring us to life. The law recognises this and makes our parents responsible for us, but nobody can go beyond that in law. Sad, but that's it. The only consolation is that death is not painful. We usually prevent human suffering, for that is what is painful in life---poverty, want, misery, disease, loneliness---worse of all mental illness. But in death we're not going to be cursed to look back to see what we're missing. Since the era of Pythagoras men have sought various means of coming back to life after death. It's no use; human life is not worth it. Life is painful, not death. I would advice everybody to try and live a good, trouble-free life, prepare for a decent burial and let his or her work in life, whatever it is, be good and serve as the lasting memory of him or her, so that those left behind could be proud of it, or probably benefit from it as well. Sadly here we are; people disparage philosophy and yet praise the distortions of relativity Minkowski propose, perhaps because only the philosopher could see that his theory was arbitrary and logically flawed.

Precisely what is wrong with the Minkowski equation of space to time is this: there is no time at all in nature, least of all general time. Yet we have something we use as time---how did that come about? This is the most important philosophical question since Plato, and Bertrand Russell put it in the clearest, lucid English words for everybody to understand, if he wants to understand it, instead of propounding mere mathematical formulas to please himself. Those who disparage philosophy should realise that even Einstein adored the title "Philosopher/Scientist"[49], and those who do not know any philosophy or how philosophers think should stick to those aspects of relativity relevant to their fields and leave the wider and deeper interpretation of the theory to philosophers. Luckily, thanks to the Russell and Whitehead examples and teachings, many other philosophers are also accomplished mathematicians. These 'accomplished' theorists have also read the judgement of Professor Eddington regarding the Minkowski proposals---i.e. to the effect that they are arbitrary and fictitious simply because they are based on imaginary time coordinates for the creation of the mathematician's imaginary infinite time forwards and backwards.

There is no general time; and if there is no general time covering the whole universe then it means there can only be something called your own 'local time', this inference is based on the fact that we know, use and call something 'time', without which we could not exist at all because we'd not know the world properly---day and night system, the seasons, knowing when it is safe to go to the farm, when to sleep and when to wake up, go to school, work, the shops, and so forth. We have time. How did we get it in the absence of a universal time? Philosophical reasoning is the greatest achievement of the human mind, not mathematics. Mathematics is like logic, somebody has to weld their pieces into a philosophic whole.

Once local time was discovered proving that time is neither universal nor fixed for all eternity, ironically, the concept of local time ceased to be important---only the idea mattered. Man was

---

[49] It is the highest intellectual honour. All human beings think as they feel, sometimes even without becoming aware of it. Only philosophers think strictly according to the facts in pursuit of the truth even when it goes against their own interests (most of the time many judges in the law courts fail to aspire to this level of objectivity; in fact, some judges are worse than criminals.) All men and women should learn a bit of the methods of philosophers however imperfectly, for it is the meaning of the highest intellectual honesty and mental acuteness.

freed from the constraints of absolute time, but then what is measured by the clock? It took a philosopher to ask that question, and it still hasn't been adequately resolved. My efforts (which are everywhere ignored because we all think there is general time) are part of the attempts to answer that crucial Russellian question. One answer is that time is variable and the variability is caused by means of unique parameters---that is to say, the elements (agents, factors, etc.) we use to create our local time vary from one place to another. Obviously the planetary orbits vary, so the years of the planets also vary---simple.

In a fragmented universe, as the Einstein theory of frames makes clear, one system of time cannot be applied with equal validity to all fragments of the universe. But then what is measured by the clock? This Russellian question will never go away. Time Dilation did not dilate any time at all. The Lorentz local time idea was important because it inspired Einstein to discover his theory of frames.[50] After that we've to discover how our own time is created---in the absence of a universal time, of course. This is what mathematicians are unable to do, but they will not accept the philosophers' solutions because they still believe that time is general, and use their esynched systems to create infinite time from the distant past to the infinite future. This is not science; it is day dreaming. And to say that it is the only means to understand relativity amounts to admitting that Einstein's ideas are still not properly understood. My advice is to think of special relativity and the two postulates underlying it. Next, you should go on to consider the theory of frames upon which special relativity is based (with

---

[50] And as Abraham Pais has suggested, this theory of frame ought to be taught in schools---especially schools of mathematics. For Time Dilation did not dilate time. it showed, crucially, that time does not exist as a general entity but that it can be created from any locality---much more serious than dilating time; but as usual, it's the use Einstein put it to that matters eternally, or philosophically, yet mathematicians do not believe that philosophy matters. I'm afraid it does.

every frame having its own natural laws, hence the relevant postulates); then move on to the metric of general relativity that is another world or segment of the universe altogether; and finally ask yourself who on earth could have conceive that idea about the divisions in the universe showing where life is possible and where it is not? He even suggested some of the natural laws too—e.g. the bending of light, etc. All these even before we come to the quantum theory! At the very least mankind should show gratitude by understanding what he said and meant, not distort his ideas with meaningless mathematics, probably dreamt up after a night of booze.

THE MERGING OF SPACE WITH TIME[51]

This is where Minkowski is reputed to have shown his mathematical genius in the interpretation of relativity, making it accessible to scientists as a whole. But of course we know now that his formula was wrong because time is not mathematical. It is something else but can be rendered mathematical in presentation---and in presentation only. Several years later I came across the report that Einstein actually never understood the theory, although craftily he praised Minkowski to placate his mathematical critics at the time. According to David Hilbert: "Every boy in the streets of Gottingen understands more about four-dimensional geometry than Einstein. Yet, in spite of that, Einstein did the work and not the mathematicians." Thank you.

I have gradually come to the conclusions that when something like this is said it means Einstein thought the theory concerned was nonsense.[52] Hence, logically what can be sustained is the concept of time proposed by Einstein (as mentioned above), in which time

---

[51] Time is linked to space (or can be merged with space), only by the 3+1 method. That is how Einstein did it. It is the only logical method, not by anybody's intricate mathematics.

[52] From A World Without Time, by Professor Palle Yourgrau, Penguin, 2007—page 6.

is seen as it is---year after year after year, or second, second, second and so forth. Bertrand Russell interpreted it as "relation between points", or that we construct time units as relation between points passing by---the years, for instance. And Professor A. N. Whitehead said it means a time system is "a sequence of non-interacting moments".[53] Of course we have already noted that the Minkowski 4-D geometry or four-dimensional space in which space and time are fused into one entity was described by Professor Eddington as arbitrary and fictitious. Scientists continue to call it artificial and yet repeat the phrase 'space-time' to mean that space and time constitute one entity as Minkowski proposed.

My belief is that this is the main reason physics is wobbling. In my view, the phrase 'space-time' cannot and should not be used to mean that space and time constitute one entity; but it can rather be used to imply that time, as relation between points, can only be had by the application of points to space, for using points involves space---so you cannot have time without space, but the two are separate, precisely as Einstein made them in the special theory of relativity---see more of my arguments below.[54] There are a lot of repetitions here, but I take the view that it is a small price to pay for clarity in a subject so mysterious that many writers are afraid to touch it! And this theory of time, as tentative and tremulously I put it forward, is nevertheless aimed at helping us to get an idea of how time is passing by; they call it the most intractable aspect of time, but to me it is all we can ever know of time. I don't think the

---

[53] The Principle of Relativity, Cambridge, 1922, Ch. IV

[54] I know nobody will pay any attention to what I say because I have published ten books repeating these ideas and nobody will even reply to my letters. But there you are. Intellectually human beings are never rational, generous or considerate except when it suits them to pretend as such. If I were writing sweet religious ideas, ah, then I could have a large following---precisely as happened to Herman Minkowski because his theory makes time universal again—i.e. prevalent in every space. My theory rather says sentience is required because somebody must be there to place the points to determine the yearly cycle which forms the basis of our time, including atomic time.

passage of time is intractable; it's time itself that's intractable because it is impossible to know what it is at all; what we call time is only how it is passing by---year after year after year. Yet the year does not help us to know what the nature of time that is driving it really is. The year is only a physical journey round the sun. Thus I conclude that all we can ever know of time is how it is passing by and never what it is.

Perhaps it is necessary to lay out the logical objection to the Minkowski proposal. I will do my best. To be frank, it is not really important. Minkowski probably wanted a piece of the Einstein fame but he chose the wrong subject; yet, sadly, everywhere we hear of the term 'space-time' used in the sense that space and time are unified into one entity, and it's the theory of Minkowski they're referring to even though every Reference Work describes space-time as artificial, and rightly so.

The Minkowski theory is purely mathematical in presentation; but every mathematical theory has to have a logical foundation, and so it is the logic of the theory that is regarded as defective. We start with his replacement of time with an equation usually referred to as 'ict equation'. This is how Einstein himself stated it: "...we must [really must?] replace the usual time coordinate t by an imaginary magnitude $\sqrt{-1}.ct$ proportional to it..." That's enough. But even more damaging is what he wrote in another section of the book (RELATIVITY): "...the world of physical phenomena which was briefly called 'world' by Minkowski is naturally four-dimensional [note that so far he's given no logical reason why it should be so but soon he did so] in the space-time sense. For it is composed of individual events, each of which is described by four numbers, namely, three co-ordinates x,y,z and a time co-ordinate, the time-value t. The 'world' is in this sense also a continuum; for to every event there are as many 'neighbouring' events (realised or at least thinkable) as we care to choose..." No wonder that David Hilbert, who must know, says Einstein did not understand or accept the Minkowski theory. He sure could not have overlooked the logical defect in the theory even if he was asleep, but at the time he needed the support of the mathematicians with 'superfluous learnedness'; so it is possible that he felt coerced, forced to used the Minkowski formula in the field equations of his general relativity, knowing that it could not in any way vitiate his proposals about the causes of gravity in the cosmos at large. I am really annoyed that physicists do not understand this; it's so simple. If I can do this without even primary school education, surely anybody can. The need is to think philosophically and they don't want to do so due to inherent professional bias against philosophy and philosophers.

The point is, the Minkowski theory is entirely based on Coordinate geometry. It had to be because it was supposed to make time geometrical; coordinate geometry has to have a premise, a logical foundation---it should be rooted in natural phenomena. The basic theoretical formula is often referred to as "The Transformation of Coordinates", or "The Lorentz Transformation". But at some stage the transformation become tenuous, even ghostly; yet one cannot logically discuss time in ghostly terms.

Time is peculiar and very strange; the most mysterious thing in the universe. It is so strange that several books, including the one mentioned above by Professor Yourgrau, called A World Without Time (note the title well), are saying it does not exist at all. Yet we have time: In sports, in work (you work according to time); in schools, in lectures, in travel, the trains and planes are all scheduled according to time. In everything we humans do time is the most important aspect. And if you are going to discuss time in the form of the transformation of coordinates you must realise that (in the nature of mathematics), at some stage the transformation will become ghostly.[55] Even Einstein used the phrase "realised or at

---

[55] It's sheer vanity to pretend that mathematicians are a special breed of people who deal

least thinkable". But it will not do. You can't use the term 'thinkable' in serious science or what he often referred to as 'logical thought', and he knew it.

I come now to the elaboration of my arguments against the theory promised above: a little theory here will help the reader to understand what I mean. Bertrand Russell noted that the merging of space and time (which Minkowski said he could make geometrical), is already evident in special relativity. It is so important that he called the Einstein theory of time perhaps his most revolutionary and far-reaching discovery, and I have rubbed the cautious term 'perhaps' to insist that it is in fact his greatest discovery far even above his new theory of gravity. Funny enough it leads to the view that there is no such thing as a universal time, yet if that is so then the Minkowski theory of time which Einstein was forced to use for general relativity is completely untenable. For if universal time does not exist then space is not the same thing as time because space is universal no matter how it is defined.

Here is the reason why. According to Russell the merging of space and time is already implied in the special theory of relativity, and I agree with him. So why did Minkowski feel that he had to do it all over again? It's because he said he could render the whole exercise mathematical or geometrical---he claimed that time, as proposed by Einstein, is naturally geometrical so that we could dispense with the 3+1 formula. You cannot blame the

---

in aspects of reality denied to the rest of us. Any gynaecologist will confirm that they're conceived and born like the rest of mankind. What sets them apart is the tendency to imagine properties contrary to existing phenomena. Many gadgets and facilities have been invented in this way; yet one of the greatest of mathematicians, Professor Eddington has warned that in an experimental science we have to discover properties not assign them from the imagination. If this advice is not taken on board and time of all subjects is subjected to the mathematician's imagination, the result will be the total collapse of physics some day.

mathematicians for falling in love with this proposal. It sounded fantastic and adorable---provided he could bring it off.

Well, was he successful? The answer must be no because everybody calls his space-time concept artificial, and although we continue to use the term 'space-time', Professor Eddington has warned everybody not to forget that it is arbitrary and fictitious, so let us look at the reasons for the Professor's advice.

Logically the merging of space and time means two things, or implies two conditions. The first is that you cannot have space without time, and the second is that you cannot have time without space.

First, the space: time is gone or going even as you traverse the most infinitesimal space; so every movement or activity must include time coordinates---a time coordinate is obligatory in everything we do, practically everything. Thus Russell pointed out in The Analysis of Matter that the tram never repeats a former journey; the reason is that, although running on the same lines, it is really in a different position because its time coordinates had changed. In physics it is even a different thing altogether.

The second point is that you cannot have time without space, because in the absence of a universal time you only get the time as relation between points—just like the way we get our basic unit of time which is the earth-year. We only can get time intervals as relation between points. The mention of points implies the inevitable use of space. After that space and time are separate entities, and that is how Einstein made them in the special theory of relativity, again as Professor Bernstein put it, "In the absence of gravity, space and time are distinct entities. In the metric of special relativity they play distinctive roles..."[56] I am therefore not in the

---

[56] This is logically straightforward. For if something is created with space and points as intervals between points (in the manner we get the year), then the resultant object is

least surprised that David Hilbert has attested that Einstein didn't even try to understand the Minkowski theory of four-dimensional continuum. The extension of the time coordinates to make it logically valid would be infinite, in other words universal, yet he had already shown that universal time did not exist!

Perhaps Minkowski saw in the whole Einstein proposal the opportunity to make a name for himself as a great mathematician[57]---but obviously not a great philosopher of physical reality; the two subjects can never be treated adequately in separation. He proceeded along a mathematical route for making space and time one entity and said so publicly---hence the word 'space-time' and some writers have said Einstein was lucky to have Minkowski on his side. It is also true that he praised his great mathematical knowledge----but not his philosophy, I am afraid. However you look at it, the absence of a universal time is blocking the way to a logically valid proposal to make space and time into one entity, and a man like Einstein couldn't have missed it.

Because Minkowski based everything on the transformation of coordinates, his proposal is not logically valid since he had to rely on imaginary time or i in the discussion of time, or his basic equation usually called ict equation: $\sqrt{-1}.ct$... But, as I keep stressing, at some stage you cannot transform coordinates indefinitely, or to infinity because there is no longer a universal time---which is what the same Einstein had discovered. You can do so in mathematics---which is what he did--- but for an elusive thing like time (elusive, basic and essentially unavoidable) at some stage you are going to have to assume that time will always be there---but in the absence of a universal time, how can that be guaranteed in logic?[58] How is the time going to be there, and there where? Or

---

independent of both space and the points---it becomes the product of the two agents of its creation.

[57] If so then he's certainly succeeded.

how? Time is always there; what we want to know is how it gets there, as Russell put it, "if cosmic time is abandoned?" I have concluded that life is time and time is life, except that the life came first; but once you're alive you exist in "a when?" situation. You cannot say somebody exist without knowing when? For nobody could be in existence without a time of his being. To live is to expend time. Paradoxically part of the mystery of time is that this time has only gradually evolved as we learned to 'construct' it out of the parameters that were always there---we always aged, but have only recently realised the reason why? The same thing applies to time: arithmetic, a theory of numbers, sentience, and the ability to count had to be learnt before time could be constructed for general use in society. For centuries, man thought time came from somewhere, say, the cosmos; but thanks to Lorentz and Einstein, we now know that it is a syndrome we construct from existing parameters. But once that is done, existence has to be interpreted in terms of time---ironical maybe, but that's the truth. With this explanation of time, it ceases to be scary and all theories that makes it mysterious or universal (like something eternally existing everywhere and the same all over, including the concept of four-dimensional continuum from our mathematical friends), have to be set against the logic of time as we experience it. I can't put it stronger than that but the reader will know what I mean, instinctively!

Thus, so long as we all agree that there is no longer a universal time, the Minkowski equation of space to time (or the creation of four-dimensional space continuum) will remain invalid in logic---but not in mathematics, which brings in an interesting contrast between mathematicians and logicians or philosophers regarding the nature of physical reality. Like their patron, Pythagoras,

---

[58] This is where philosophy comes in yet scientists disparage philosophy; they're only partially right. Not all philosophers write logical mysticism that seeks to put an end to physics as Wittgenstein did---according to Russell.

Mathematicians believe that what is logically valid in mathematical deductions is or can be encountered in physical reality. That is called 'inference', but it's no longer applicable. Russell and Whitehead have shown that the world of sense is a construction not an inference.[59]

Conversely, philosophers believe or think they know that what is proposed in mathematics must be discovered by philosophers and theoretical physics or is mere speculation, if you want to be friendly and charitable, or mere dreams from pure mathematicians as they are wont to do, if you want to call a spade a spade. Some very unkind things have been voiced about the habit of pure mathematicians, but then we should remember that even Newton said he's afraid of them---several of my own technical papers have been rejected by mathematicians because of my criticism of Minkowski, yet we are now coming to the conclusion that his equation of space to time is logically untenable. Who knows what progress we could have made if this was recognised earlier?

As stated above, it is possible that the Minkowski theory is causing part or even most of the problems facing physics at the moment; and we ought to realise that physics is now so complex and vast that no one scholar can stand up against it. With all mathematicians inevitably backing the physics establishment, it's impossible. Yet even Eddington called the Minkowski proposal arbitrary, so did Russell, and now we know that the great man himself, Albert incomparable Einstein, also did not accepted it--- not even bothering to understand it I accept this as true, since we

---

[59] Sir Karl Popper is wrong to say he's not proud to be called a philosopher due to the silliness displayed by some philosophers like Wittgenstein. A great deal has been accomplished since Russell and Einstein, to which he Popper himself has even made substantial contributions---the problem is persuading scientists to look at what philosophers are doing since Einstein's revolution. Even if the answers elude them the correct line of thought will be indicated. But I agree with the scientists that the way the mad thinkers are praised by the big universities does not boost confidence.

know that he intensely disliked the mathematical interpretation of his ideas.

It is a serious matter and I hope physicists will reconsider their slavish use of the term 'space-time' as meaning time and space constitute one entity. Eddington called the Minkowski proposal arbitrary and fictitious but added, "...Such a mesh-system is of great utility and convenience in describing phenomena, and we shall continue to employ it but we must endeavour not to lose sight of its fictitious and arbitrary nature." I believe Eddington was partly responsible for the mistake over Minkowski, for this is a very learned ('British') way of committing a literary and intellectual crime of the century---urging scholars to continue to use a theory so roundly condemned by himself? And his offence is made worse by the cunning manner he pointedly refused to mention Minkowski by name. Nobody would mention this, but there were two theorists working on relativity---Einstein the originator and Minkowski the interpreter. This was known all over the scientific world. Professor A.N Whitehead pointed out in The Principle of Relativity that we owed relativity to both. So it didn't take me long to figure out that it wasn't Einstein's theory Eddington was describing but that of the other contributor to relativity!

Technically this is all because you can never extend the transformation of coordinates to infinity; so our very able physicists should understand this anomaly at the heart of physics. When this has something to do with time, you have got to define it first and nobody can ever define time.[60] What we call the definition

---

[60] The nearest logically conceivable (or acceptable) definition is the one given by Professor Richard Feynman. Time, he said, is how long we wait. Actually time is from when to when, how long we wait, growth, decay, ebb and flow, any motion, inertial and momentum, and just silently being there minding your own business! The problem is how we get these, all of them or some of them into the clock ticking time away in the absence of universal time? That brought in theory, some logical, philosophical, religious and

of time is only how it is passing by and never what it really is. In the past this was not generally appreciated, as everybody continued to treat time as if it were a universal entity; but nobody can hide behind that lame excuse any more.

Having dismissed the other three systems of time as logically flawed, even though people still believe in them[61], we are left with the true and pragmatic concept of time (the way we get the years), as one in which we 'construct' time, as Russell put it. That is how we can logically acquire the 'logic of time in the universe', showing how time (as non-interacting moments or time in units created as relation between points) is possible in any part of the universe, namely we simply do not know what time is, but can use repetitive cycles (like the year) to give us numerical quantities of how it is passing by, ten orbits of the sun means ten years have passed by and gone; and that, of course, can be mechanised into a clock as we are doing; it is the theoretical explanation of what we already have, know and use as time---no matter how we got it. Even religion played a part.[62] In the words of Bertrand Russell (to quote him in full):"It seems that the one all-embracing time is a construction, like the one all-embracing space. Physics itself has become conscious of this fact through the discussions connected with relativity."[63] The rest is a simple matter of straightforward deductions. However, it has to be acknowledged that human beings are extremely fragile both in mind and body (psychologically and physically.)

---

mythical. But science, upon which we rely, demands what can be proved so that we could feel safe to rely on it. Definitions of time must not fail this test, but the Minkowski formula did. I hope my explanation is satisfactory.

[61] Because of the fear of death as the end, ideas about time and time travel that imply coming back to life are popular.

[62] See the essay on Time before and Time after Einstein below.

[63] From Mysticism and Logic, Ch. viii (x).

We should not really be here at all and we know that many factors have contributed to our survival, including religion. Some obvious myths may be mocked today, and certainly not wise to continue to observe them---human sacrifice, for instance. Yet in the dark old days they might have contributed to our survival. The same story underlies the having of time, since it is not cosmically imposed. That is how important the Einstein notion of time as a purely secular or psychological entity happens to be. Religion, astronomy and astrology, mechanics, mathematics, the ability to count, physical suppositions or concepts (as physics) and sheer human ingenuity or sentience, all contributed to human beings acquiring the concept of time; they are all part of the logic of time in any part of the universe. Sentience was particularly important because somebody had to be there to set the points for the yearly cycle. Any 'Beings' anywhere will face the same problem. This is what we can logically trace about time overall. And I have to emphasise that, despite the Minkowski contribution, which I call 'a distortion of relativity', Einstein in fact made space and time separate in the special theory of relativity.

There is something to be said for the notion that time does not exist. I have even written a small book about this problem, entitled Why Time is not a natural Phenomenon. But there is something we know and use as time. So time is real enough, but to be really philosophically or logically accurate, we should regard time as a device we have invented to guide our actions and that it does not exist naturally in nature outside the human mind. Thus time can be defined as: "existence" and does not move; what is moving up and down are the cycles we use to regulate our actions and which we call 'time'! This may be a paradox but no longer mysterious since Einstein's demolition of Absolute Time. So the phrase 'passage of existence' is almost right, except that time does not move, only the cycles we use for time are moving and the passage of existence is chemical not temporal.

CHAPTER FOUR

THE FOUR AGENTS THAT CAUSE WHAT WE EXPERIENCE AS TIME.

One universal constant of 'Being' is that everything in nature is caused, whether the causative agent and processes are ordinarily experienced by man or not.[64] It is part of the duty of professional thinkers to investigate the causes of things and events as their full-time job. We are in the habit of calling thinkers philosophers; but there are other thinkers who are not philosophers---scientists, mathematicians, logicians, all of these work, in the main, on our problems, trying to solve them for us, and one thing they all agree on is that everything has to have a cause. What we call 'research' is the act of searching for them.

For ever trying to create problems for science, the religions go further to insist that man's life must also have been caused, and as we cause human births, somebody must have set the wheel in

---

[64] Even the most extraordinary organ in the universe---the human brain itself---could not have come into existence unless the elements that created it had the capacity, in combination, to create a brain.

motion for giving birth to humankind or we could not have come to exist from thin air. From this they deduce that God must exist, and I would advise people to desist from trying to condemn or defeat them in arguments over this because they would rather die than accept any other (scientific, logical) explanation for the being of human life, since by this same doctrine death is only a renewal of life, and yet, despite their pretences, everybody is afraid of death; and of course nobody should try to deny people the right to believe what they will because some people couldn't live without their religious beliefs.[65] Another thing we all agree on is that the causative agents may not be immediately apparent, their effects may delay through inertia or chemical processes, but causes are the agents (creators) of everything, gases, solids, atom, chemistry, et al. Some things happen to cause events just for being there without knowing what they're creating.

Going by the philosophy of causes in nature, so far, four natural agents, factors or conditions have been identified; it does not really matter much how they're described. The reality is that beginning from atoms things are caused by combinations of atoms, whether we as human beings do experience them or not, particularly quantum mechanical causes that are buried deep in objects which were never suspected until the dawn of the Einstein Age.

Many people believe that the Einstein Age began with the theories of relativity. In fact, his second greatest discovery was the causes of the photoelectric effect; the first was his theory of time, particularly the discovery that the universe is fragmented and exists by means of different parameters so one system of time cannot be applicable to all of them. Wrongly the general theory of relativity is

---

[65] One rule that should never be broken is that once a person is born he or she is entitled to live his or her life to the full---nobody knows how he or she got here so nobody has any right to take life away from anybody.

regarded as his greatest just because of the gravity aspect of it. One can never understand why human beings are so concerned about what happens in interstellar space---it's so far away and not controlled under anybody's conscious direction at all. The British in particular (always sulking and moaning, cursing everybody for the birth of Einstein to come and dethrone Newton), are so in sensed about the new theory of gravity.

Ordinarily people think insights about the cosmos are more valuable than insights about the little pieces of matter that actually make-up the mass of our bodies. But to discover that the world of visible matter is actually controlled by the movements of the invisible quanta, is my number one discovery of all time, (the Nobel Committee also agreed), followed by the idea that time itself is created, in the words of Russell, "constructed" by man not imposed by God, and bound to vary from place to place. Gravity comes third.

As I keep repeating, some of these causes of events and time can be used to show only how much time is passing and never what it is[66]---that is this writer's special position.

It may sound like fantasy (even treachery to some philosophers), but I think the religions believe the problem of time

---

[66] I have several theories or philosophies of time: one is that the passage of time is all we can ever find out about time therefore we need no theories to account for the passage of time---the years replicate or increase in numbers to pass by; they need no arrows to point the way. That is a human concept and not applicable in the universe. The second is that time that is derived from points, like the earth year, cannot march; it is bound to be discrete. The third is that discrete time does not run through the universe; so history is not the cause of time but of events; The fourth is that discrete time cannot curve so the concept of 'Curved Space-time' is logically untenable, and following form that, the fifth is that time travel is not only nonsense but a disruptive nonsense. The sixth is that you cannot say time does not exist; explaining what it is proving rather difficult, but surely we use time in all activities---travel, work, schools, sports, etc. The seventh is that all these logical ideas about time makes it secular and the religious interpretations of time is sheer humbug.

can never be solved, and that after Darwin time is what makes them believe still that God exists and that he created us. It's no use citing biophysics and all the rest of it. Time and life go together. We know life because we are living it. We can't know the nature of time though we are using it. It's therefore open to a variety of interpretations. Choosing the logical or scientific interpretations does not preclude others from opting for the religious view. We have to remember that when we use physical cycles, like the yearly cycle, to track time, what we are doing in fact is showing or recording the rate of the passage of mere physical cycles and never the time itself----whatever it may be.

The reason, of course, is that the physical cycles are not time; and we don't even know how long the year is in terms of duration in the mind. They give us just how many times (orbits) the earth has recorded. We count these cycles and call them 'years' or whatever. So that for mankind, ten orbits of the sun is ten years; and one would have deteriorating physique to show for it. Is that time? Certainly not. We count them as the rate of time but they are mere physical cycles; so it means we count the physical cycles as the rate of the passage of ten years. There is no way that these physical cycles could be the real nature of time, only how it is passing by. The division of the year into months and so forth means nothing---we still cannot define one unit of time. For instance, how long is one second in logic? One minute? One hour? They don't have meaning because we can never know what duration means, and counting the years is just a matter of counting cycles, tapping the finger is the same thing. It proves nothing.[67] Perhaps we age as the passage of time. Ah, ten years will make everybody age in some way---more or less. So can we regard ageing as the basis of time? How much do we age as an indication of the passage of one

---

[67] It's from such arguments some writers insist that time does not exist. My reply is that it is not an honourable stance for philosophers to dismiss something they use daily as a myth. Time is not only in real existence but is oppressively unavoidable.

year? There is no clear landmark, and also it varies from person to person. One can see that the religions are no fools. I believe totally in secular time; but those who do not cannot be dismissed as easily as that. And since the religions have money and power they want to keep their privileges with as much mysticism and doubts about nature as possible. That's what they are doing. It's not fair, but there you are. Many things in life are unfair to the losers; but you can't blame wealthy and powerful people for trying to preserve their advantages in life. Life is so nasty that anybody who has any advantages to help him or her live comfortably should seek to preserve it. After all, self preservation is the brain's most powerful, oppressive force on people. You can't avoid it and live.

In any case, here are the purely secular factors or conditions (strictly speaking they're confidence tricks.)[68] We use them for reckoning time for general use either individually or in various combinations. Obviously time is a construction; as such we use the parameters we assign to nature or recognise in our environments to construct it, and therefore bound to be different on other planets. But time anywhere else in the universe will have been 'constructed' the same way—by the uses of either one or many of the following agents. The Russellian notion that time is a construction seems the most logical explanation of time under relativity. Not that time cannot exist under the Einstein proposal---that was an expression of intellectual laziness---but that it is constructed by the human brain. So it's the brain we have to decipher atom by atom, if we can! Not only about time, why is it able to probe the entire cosmos? The whole idea of being in existence, of being human, centres on the brain's extraordinary incisive powers. What is the purpose anyway, since we are so ephemeral and infinitesimal? For me it shows the infinite generation and regeneration, death and rebirth round and round indefinitely. Pythagoras was nearly right, except that the

---

[68] They work out of necessity.

same material does not last; atoms disintegrate. Otherwise what is the meaning of regeneration? So, for me, death is sadly the end. Traced from the nature of time, this is philosophy at its deepest level. That's how important I find time to be---even then we can only know how it's passing by and never what it really is. Perhaps the greatest mystery in all nature is time, and this is how it is caused from my point of view:-.

(a) Chemistry (physical and organic);

(b) Quantum mechanics;

(c) Motion;

(d) Inertia.

I will now give brief comments about how these agents can cause what we experience as time or a period of waiting, always by means of the brain, the real mystery in all nature.

First, chemistry. The most obvious example of chemistry or chemical processes causing time as a period of waiting is human gestation or any gestation at all; though human pregnancy is what concerns us most. The normal period is nine months. However nine months constitute the time for the growth of a fertilised egg (the foetus) to the point of birth or a normal baby---ready to face the world, with what luck nobody can tell! We call the waiting period 'time' but in chemistry it is a long process of growth, the conversion of matter from one form to another.

Next, the quantum debate. Quantum mechanics are not perceptible, but they cause a period of waiting, though most likely to be very short---it is still time.

The third factor is motion. Motion, obviously, can be seen as time. It takes time to move from A to B. For instance when a sportsman throws a ball into the field of play, it takes time to travel to reach the players in the field. Motion of any kind can show time going; but it is never the real time in force; real time, as I have argued, is never known; only its rate of passage is recorded by the clock together with its psychological effects as the sense of duration. In other words, *if real time is unknown, then the cycles we use to reckon time constitute the time in so far as we are concerned.* This is what is taking mankind so long to grasp since the abolition of universal time, so much so that nobody will even look at what I write! One agent who did write back said: "I am sorry but not being a professional mathematician or philosopher... I am afraid I cannot figure out why this is important or what it's about..." Brilliant! I am also afraid that it is important because scientific progress (chemistry, medicine, physics, electricity and all that we need for the proper and safe control of our environment) depend on the scientific understanding of time. The problem is that he holds the money, the power and the publicity levers and he does not understand me, or the fact that we need to seek the real nature of things so as to be able to control or manipulate them to our advantage. Even Einstein admitted that he was able to complete the special theory of relativity a few weeks after he got his insight that the Lorentz t^1 is 'time, pure and simply'. Lorentz also said he thought he was unable to discover special relativity because he failed to take his discovery about time as important---and dare I mention that the practical benefits of relativity are too numerous to mention? Without time we can never tract dangerous asteroids.

To go back to our story about the clock, the units produced by the clock are obtained elsewhere and programmed into it for reproduction. The ticking of any clock is not original; it's meticulously conditioned to do it precisely in the form that is done. The units of time are deliberately created (and mathematically divided) to accord with a full orbit of the sun, since we then have to start the orbit of another year. Often people equate motion to the rate of time. Such erroneous statements about the metaphysics of time are all over the place. Every time you mention time you're making a metaphysical statement about the world. It may seem familiar but you could not form such ideas if the earth did not exist. And we do it all the time. Even as I write a newspaper report (The Times, 16/9/13) is repeating the mistake.[69] They wrote: "Flies and small children may have something in common: the ability to slow down time...by seeing the world in slow motion..." This equates motion to time but that is totally wrong. Time, of course, can be based on motion; and every motion can be seen as 'time going'. The problem is how much time? For this reason not every motion is metaphysical time. It is some time going but how much in real time? That can only come from the true nature of time, and that is obtained from the breakdown of the yearly cycle before being programmed into the clock for reproduction in specific units to accord with a full orbit of the sun. It's only when you understand this will you come anywhere near the metaphysical nature of time.

Otherwise if any motion is time then the faster the motion the faster the time and vice versa. That could not give us a stable environment; for all activities are controlled by time. This is an additional evidence that motion (any motion) is not time per se, and all statements to the effect that time is mere motion should be regarded as mistaken. For they go on to claim that gravity 'slows' time or speeds it up. We can only do that by affecting the motions of the earth. Real time, again, is unknown. What we do is use mere cyclical motion to show how much time is passing. As we can only know how time is passing and never the real thing. We are dealing with something almost like quantum mechanics: we cannot demonstrate the entire range of quantum action within matter or atoms that result in chemical changes at the visual level. Yet through the theory of QED we know that they occur and that life depends on them. We couldn't exist without it.

---

[69] Frankly I am not surprised that my books are not read. Everybody thinks he or she knows what time is---even flies can slow it down, really?

I would combine the effects of motion and inertia into one causative agent. Motion and inertia combine to cause activities in stellar bodies. If a planet is moving to or being sucked into a black hole, it could take centuries, or at the very least long enough time for us to get married, raise children, grow old and die; even for several generations to do the same thing; or fight wars; conquer other nations; establish civilizations, lose them, and start all over again!

All of these either singly or in combinations can cause time interpreted as a period of waiting---not as 'time allowed' by 'The Creator'. It is an understandable mistake caused by philosophical ignorance to claim that the periods caused by all or any of these alone or in combination is not time like the time assumed to have been bestowed by the mythical Creator of the discredited religions---which means all of them.

Here is my philosophy as regards the act of worshipping deities. The prescriptions (liturgy, prayers, incantations, etc.) are revealed by ordinary people. They call them revelations. I call them dreams. In life you have to get education to survive, even as a child you have to be taught how to use the toilet to survive. So education is crucial. But the basic thing we learn in our education is how to reason, how, for example, to know that certain objects that resemble food should not be put in the mouth---how to form ideas about things. First you identify things, observe them, establish as far as possible what they really are (e.g. dogs are different from tigers, so that you would not go and play with tigers), and act accordingly, or according to tradition and safety to yourself and others. This process is called logical reasoning. Without it you would be fantasising about the world and the things in it, but won't last long because there are so many dangers from snakes, tigers, people and inanimate objects that can very easily put an immediate end to life. Given this precarious conditions of human life on earth (where we are forced to live without knowing why or how we got here), it is unwise to rely on somebody's mere dreams as the prescription to regulate the course of human life, especially en-mass. We know it is unwise because it has led to numerous accidents and ruin. Science gradually evolved after centuries of trial and error to help us identify the nature of things and teach us how to deal with them in safety. No religion can do that and therefore all the religions are dangerous---but there is a fear in people that unless they hear something sweet about life and after life they'd not live. Unfortunately delightful religious sermons are not associated with safety and progress; it could even be the reverse, as some people dream that the gods require human sacrifice and come to take you away. We have learnt through bitter experience that some people are basically evil and infiltrate the religions just to wreak havoc with their fellow human beings' lives, no matter that they utter agreeable sermons---it is a deliberate ploy to seduce people and ruin their lives. For this reason worship of the gods dreamt up by others is unwise and very, very dangerous. But what about our own dreams, then? Again, dreams are outside logic and therefore unreliable; it is better to stick to what has been rationally examined before you trust your life to it.

We are talking here about secular time: after Lorentz and Einstein found that absolute, general or universal time does not exist and running all through the cosmos and the same everywhere, from the past to the present and the future, it became immediately necessary to investigate how our own time began; and this line of thought has gradually proved irresistible. It is logically the most convincing reason for having time.[70] It should be emphasised

---

[70] I believe the bogey of past, present and future was spawned by the study of history as a story from the past, giving rise to the present and carrying on to the future. People think of the process as one of 'time marching on'. It is precisely as they describe it but it is not time that is marching on---it is the events. Suppose you borrow money from your bank, it is the money he will chase all the way from the past to your present circumstance and

strongly that time was not only thought to be fixed but that it was the same everywhere. That is the reason we think Einstein, being so clever, had a brain wave that if it is changeable then it could not be the same everywhere and so on. This may sound elementary, even tautological, but every little thing about time takes a genius to clarify. Nothing is more mysterious or linguistically intractable.

THE ORDER OF TIME SEEN AS A MATTER OF ARITHMETIC

The human mind craves order and cause, so we think of the order and direction of time—by who, or imposed by whom? For God's sake we've just been liberated and freed from the bondage of time's absolutism only to be confronted with its restrictive order and direction. I think it's all the fault of the brain. The brain requires order and direction due to the way it's put together. We imagine that several compounds cobbled together accidentally to create the brain.[71] With each component searching for 'complement', the probing tendency in the brain was established as part of its basic structure. This may be mere speculation, yes; but that is how we've gained all the knowledge we have.

Even then, there is some tenuous evidence for this because we can see how it grows in the foetus as the host (the woman's womb) supplied it with the necessary compounds and chemistry, and when complete, begins to take over the foetus and cause its growth to the point of birth; after which it is fed externally to direct the body to the point of death. Originally it must have grown compound by compound. The process probably was that one compound and another joined up accidentally. Then the quest for 'complements',

---

your future earnings, not time. Bank managers are notoriously careless about time, not the money. He'd gladly extend the time of the loan if it will bring him more money in the future!

[71] Recently scientists have created a human brain in laboratory by the same methods.

the need for order and the sense of waiting resulted, caused by what I can only describe as 'chemical hunger'. Most of what is written here cannot be proved; but I think it may come close to what actually happened. For something must have happened to cause the generation of the brain out of inanimate matter, elements or compounds. This is an attempt by ordinary human beings to trace the physical origins of the human brain. It's not rocket science, as they say. The important thing is that it's not religion or fantasy either, for the life itself grew the same way: elements formed the original, egg or sperm to create the basic primeval amoeba that replicated till it grew to be a sustainable organism. It is also evident that the brain's demands on the human body (including forcing us to have and endure the sense of waiting which we know as time), are built into the brain's structure; and we imagine that it could have come from the protracted, chemical processing over many centuries involved in the brain's creation out of the elements. This word 'element' is used instead of saying atoms, but it is to be understood that its ultimate constituents are believed to be atoms. This, of course, is the language of science. It is conceded that not everybody believes in science; what is evident, however, is that nobody can live without science. So some of us abide by it on pain of death. Otherwise it's not illegal to believe in anything that does no harm to others.

To continue with our story of how the brain probably evolved, we can imagine that when the required compound for complement arrived, more chemical tentacles were created---for the brain is more than a million times more complex than the biggest computer ever created by man---more material were needed for a complete organism to reach an independent and sustainable whole without constantly casting around for more materials. Hunger, the sense of waiting, yearning, cravings, the need for reason to explain causes and the need for completion to calm the increasing number of chemical tentacles requiring 'soothing' (roughly, it is thought), caused the brain to come to exist with these tendencies built into it. One of them is time, or the sense of waiting (waiting for the require complements), which internally we know as duration. Another is the craving for order and causes. They do not exist in the universe outside the human mind.

Otherwise by whose order or direction is the putative arrow of time following? It may very well be that the order and direction of time are misleading concepts, precisely the manner we get the years and the centuries. Time does not seem to move physically only mathematically; you can easily alter ten year to twenty on paper or through mathematics; the physical aspect of time is external and comes from the motions of the earth. After that human ingenuity took over to create concepts of time in the mind to accord with the motions of the earth, which may be regarded as the birth or the cause of the birth of our whole concept of time. Since the earth's motions are repetitive, the units of time spawned by them are also digital, and we applied arithmetic to them. Thus, unlike our primitive ancestors' practice of keeping time with charcoal marks on the wall to indicate the number of days, weeks, and months gone by, as the passage of time, modern man has the use of theory to simplify things for him---after that the mystery-makers took over, beginning with Pythagoras.

Anyway, because of all this, we now know that time is produced in units and the units multiply for it to advance or move on. Is that still disputed, with the example of the yearly cycles being so clearly demonstrated? But it's essentially a matter of arithmetic: if the year is one unit of time then it replicates to become two, there, four, five, six and so forth all the way to the centuries. Is that still disputed by the mystery (or mischief) makers? Yet it confirms Russell's idea that we actually do construct our own time probably through the brain's cravings for order; so the order of time is in the mind not in the world out there.

Sentience, a theory of numbers, the ability to count and points are required to fulfil this order. Hence space is involved, since the use of points implies space for the creation of theoretical time based on mathematics or arithmetic. We can still call it 'space-time' but only in the sense that time is or can only be created in association with space; and its passage, too, is the same as its ordered progression---i.e. through the procession of its units, a matter of arithmetic. The years increase in numbers to pass by, without any direction. For one thing we've been able to establish clearly is that things are created continually in the universe through the accidental combinations of elements, compounds, atoms, et al. Even the brain created itself and dies to prove that it is not a permanent entity, just a passing thing produced through accidental causes.

All the same, some writers have made the order of time the pivot of their own interpretation of time, even though the definition of time as the 'irreversible general passage of existence' was meant to refer to a time system imposed on us to run generally in one direction through the entire universe and the same everywhere, with the condition that we are all moving in tandem to the Day of Judgement as the end of time---the biggest folly in the human mind's suppositions, incredibly, illogical and totally without foundation, just part of the Christians' meaningless fables that we have wisely rejected for over a century ago.

After its rejection the order of time should have been seen as a purely logical matter easily resolved with arithmetic. Let me explain. The phrase and particularly the word 'irreversible' mislead people into thinking that we are all moving irreversibly with time (plants, animals, entropy, vegetation, the seas, rivers and streams, et al) to a predetermined end or destination. This has spawned numerous legends, theories and beliefs mistakenly; yet the order of time merely means the arithmetical order or progression of time's units---the years, for instance; exactly like counting the years from one to a century. And one year is also pared down to the seconds, that should reach a certain number (counted progressively in arithmetic), to coincide with a complete orbit of the sun---and start again. Because of the restart, the number of units counted has to be exactly correct. The phrase does not mean an irreversible passage of time that we cannot interfere with, but irreversibly leading us to doom. This is what scientists have magnified into the theory of entropy's irresistible march to the death of all activity. The theory of time has accumulated thousands of myths, fables, fantasies, legends, lies, religious beliefs and even mischief. I am quite sure several sacrifices have been carried out with human beings about time to placate the Gods. Yet the order of time which some writers regard as an insoluble problem implying divine influence is nothing more than the progressive counting of the units of time, after all that is how we get the centuries.

The order of timer and the supposed irreversible passage of time should have been eliminated from the debate as soon as we realised that time is discrete as intervals or time units between points; for obviously we do not all move as such;[72] the Heaven is nowhere; God is supposed to be dead; and existence is not even uniform, neither do we all move in one direction. In quantum theory the direction of motion is not even known. Some things are stationary, others are moving in reverse, and others are moving in any way they prefer. Even in the solar system presumably controlled by the sun's gravitational attractions, not all the planets move in one direction. In any case, once we found that time is variable, it was unwise to insist that we are under the command of

---

[72] The order of time and the progression of the units of time from one year to a century amounts to the same idea, solving the problem of the advance or passage of time. There is no problem of the order of time to be resolved, if the time is discrete and not running through the cosmos in some kind of continuous thread from the past to the present and on to the future. In our ignorant past we thought the order of time was also fixed. But once time is seen as discrete the order becomes a matter of counting the units.

one kind of time moving in one kind of direction to a solitary mythical destination. We now know that time, existence and motion are all variable.

In spite of all this, everybody comes to the study of time with his or her own agenda without reference to logical truth because religion and ancient traditions have so conditioned our minds that we all think we know what time is. Einstein alone showed (he did not just say it; he proved it by experiments) that we are all wrong, and Bertrand Russell not only agreed with him but said that his theory of time was, perhaps, his greatest achievement. For me there is no doubt about it. General relativity is not Einstein's greatest achievement. It's not even his second. It's his third after time and the quantum theory. I am of the opinion that what we find in interstellar space is of secondary importance to what we find here on earth; for even if we discover a body on a collision course it is what we find on earth that could be used to neutralise it. In most cases interstellar knowledge is mere intellectual pastime for selfish, psychological satisfaction. Something much more like an ordinary labour of love. The near star mentioned above may explode and destroy not only life on earth but the earth itself. But what can cosmologist do about it? ---absolutely nothing; as I have said, except to give us sufficient warning to go and join Richard Branson in his special plane, but even then where to?

I know that human vanity is bigger than the sky, but obviously the universe is just too big and complex for us to worry about putting anything right up there. The causes of its nature and activities can hardly be more important than the price of bread here on earth. No matter what we do or believe no course of action by human beings (as insignificant as we are) can make any difference (other than, perhaps, diverting or destroying asteroids on a collision course.)

Although not a believer, I sympathise (somewhat) with believers about the purpose of human existence---what for? Man is so insignificant. Even the planet itself is just a tiny dot soon to end up in a black hole and burn out of existence; yet human intelligence is so far-reaching, so inquisitive, so hungry for knowledge of the cosmos probing, probing, and probing, to no avail. For all his insight about the cosmos, Einstein died, decomposed and disappeared out of existence altogether. But his insight and discoveries about the world we live in are in daily use for the benefit of mankind. They certainly are more important than knowing about black holes. So while the human brain's creations can last, we the creators, the bearers of the brains, could die easily. Men are so fragile. The religious people are not that stupid---there is a real problem with human life on earth; there is human yearning for salvation; we just do not want to come into the world to die in misery---what is the point of that? The religions want to claw at the tiniest straw that could be interpreted as giving meaning to the senseless thing called life. The alternative is sheer emptiness, misery and early death for no sensible purpose whatsoever. Thus, despite my basic irreligious beliefs, I still think religious tolerance is the beginning of genuine humanitarian wisdom---"Remember your humanity and forget the rest", was the last advice Bertrand Russell gave us. And Russell, for the ignorant, was not just another human being. He was the world's greatest philosopher at the time, and most likely as clever as Aristotle.

But as the result of the new Einstein theory of time, we now know that there is truly no longer a universal time as Bertrand Russell put it in his book ABC of Relativity; so we have got to search for the mechanism of time or how we get our time--- or what we call time. Atomic time is included in earth time, as part of the time we have created with the earth's motions. It is often assumed by some religious scientists that atomic time constitutes a cast-iron proof that time exists in the cosmos and can be measured in many different ways. In fact, atomic time is not different from earth time. The cycles, pulses or oscillations are merely shorter than the long orbit of the sun, and, in any case, they have always to be related to the second to make sense. For this reason atomic time is still part of earth time. One can even tap the finger. It is the same thing---something we can count as the rate of the passage of time is all we can have for the reckoning of time. It is through sheer hard work and amazing human ingenuity (mostly by the mathematicians), that we have 'constructed' what we call time to guide our activities; and since this time is based on the earth's conditions, not all of which are conducive (or compatible) to living without sensible controls, our time is strictly tuned to show us the safe periods and areas of the world's conditions and environments we can negotiate in safety.

Otherwise there is no time in the cosmos at large. Our time is unique; there is no doubt about it. For it is created with the unique parameters of the earth. If there is natural time behind the parameters we simply cannot know it, because that is not what we know as 'our time'. What we call time, or our time, is 'constructed' from the parameters as their effects only---counting physical cycles as 'years' is not time. Perhaps they are the effects caused by natural time behind the parameters, but we simply do not know.

As already mentioned, Einstein divided the universe into two, as we know: they are the metric of general relativity where there is no tolerable conditions for human lives, and the metric of special relativity where life is feasible. In this home of ours there can be time, as we suppose that in similar homes elsewhere in the cosmos time will be thinkable. Where there is life, there will be time. That is part of the logic of time in the universe. Every time system can only be based on physical parameters; and they are all different one from another. Some or most of these parameters are present in all the segments of the universe; otherwise there is no time in the universe. To have time you've got to have the intelligence to construct one out of the components of the relevant parameters.

The conundrum is this: on the one hand, we think there is something called time naturally moving on by means of the factors or agents mentioned above; but if true, what then is moving this time on? On the other, it would appear that, like the brain emerging from nowhere and seizing control of everything in sight till its own demise, human ingenuity has created what we call time out of the natural features (or parameters) we find in the universe. It appears these parameters or features, being mere physical materials, would know nothing about time as the sense of duration in our minds.

Being the greatest philosopher of the period under discussion, Russell asked the most important question about time, enough to redeem philosophers' reputation after their condemnation by Karl Popper. When Lorentz and Einstein showed that absolute, fixed or general time permeating the whole cosmos (and the same everywhere) does not exist, Russell asked, what then is measured by the clock? Frankly, apart from the orbits of the sun, there is nothing (unless you can tap your fingers continually as the rates of the passage of time). Hence the thought of secular time, which is the same as counting the days and nights as the passage of time.

A careful analysis of the Russell question gives a perfectly logical explanation of all aspects of time as a secular entity 'constructed' for use on this inertial frame, and even then only capable of showing how much time is passing and never what it is---provided one can ignore the billion or so myths about time.[73] We use cyclical or regular motions to give us time--but they are physical, so they can only show how much physical cycles are passing (have passed or will come to pass, e.g. as 'years'), and we

---

[73] This answer, though very theoretical and academic is not much different from counting the days as the passage of time---or even tapping the finger to indicate the passage of the seconds.

use them as time periods to plan all activities: ten hours means it is time to do so-and-so for so much hours, etc. What the real time 'is' we can never find out, only how many cycles (being the years) of it have passed or are passing. As I have said, my guess is that time is a combination of chemistry and motion (especially repetitive motions) and sentience; none of these on its own can be mechanised into a clock as time, but in combination they can give one 'a period of waiting' (especially in chemistry), which is time. Sentience is required because somebody must be there to set the points and count the orbits of the sun as years or there will be no years and seconds derived as fractions of the year. Until we were wise enough to do so, man had no time and lived like a beast of the forests. Just look at the story of the evolution of the clock since we came down from the trees.

Of late many books have come on the market dealing with time travel, and it seems they have been very successful, so successful that publishers fail to notice the contradictions in their theories. I am ashamed by this trend in publishing (where even convicted murderers are given millions to tell their stories).

For instance, in A World Without Time (Penguin, 2007, already cited) Professor Palle Yourgrau states categorically that Kurt Gödel has "proved that in any universe ruled by the Theory of Relativity, time simply cannot exist..." At another page he says it has been proved that time travel is "a scientific possibility", and continues, like the rest of us, to live in Einstein's Special Relativity frame where he says "time simply cannot exist". It is difficult to see the logic in saying anybody who is not a magician can travel by something that does not exist. It's appropriate that the book is called A World Without Time. To me it's a world of fantasy, for this world we live in certainly has time: all workers go to work by time.

One implication of all these contradictions is that the phrase 'Space-time' may be sensible, succinct and cute to some scientists and those mathematicians who want to reject the 3+1 formula for representing physical reality in space, but it's not actually true of the physical world, unless it means time can only be gained through the application of points to space, so that we get time units (or time intervals) as relation between points, like the years, not in the sense that space and time constitute one entity---just to avoid use of the 3+1 formula--- so that as space curved in general relativity it would take time with it, as 'curved space-time' for you to meet your grandparents even before they were married, just to justify religious sermons about time travel being 'a scientific possibility'.[74]

It is obvious that the universe has no time as an oppressively unavoidable order of action, as we have on earth, that is why we get problems with the quantum and other sub-atomic particles.[75] Neils Bohr said whoever is not shocked by the quantum theory has not understood it. That statement should be turned on its head, namely whoever is shocked by the quantum's strange behaviour has not understood it---he or she does not realise that the quantum

---

[74] I am trying to reveal what I know or suspect of the innermost yearnings of many pure mathematicians. We know that they tend to be mystical in their thoughts; but it is not right to dress-up their secret yearnings as objective truth. That is not how we acquired the dependable knowledge to go to the moon and back, or cure some of the dangerous diseases that plague human life. It's not fair. We spend billions to keep these mathematicians in the style of life conducive with calm and serious mental exertions only for them to make spurious claims that time travel is a scientific possibility just to satisfy their own secret mental cravings. All those who believe that time travel is possible are free to leave the earth in the next hour---an hour is long enough for them to pack their stuffs. Professor Eddington's advice was that in an experimental science we have to discover properties not assign them. The fact that he gave this advice in the Introduction of his Mathematical Theory of Relativity had added poignancy.

[75] These existed billions of years before visual objects were actually caused by their own interactions to come into being. How, therefore, could they copy the actions of visual objects? They don't know them! This, of course, is speculation, but the whole of solid science grew out of such speculations. Even Wittgenstein's logical mysticism is hailed by Oxbridge as 'great philosophy'.

alone (without conscious direction as in LASER) is not subject to time or what we call human time, 'constructed', out of matter after the quantum came to be in existence. I would advice that we look at the quantum carefully. From what we know of its nature, we understand that it would have been there (in a strange sort of existence we can never imagine) long before their interactions caused objects to come to be in existence. It can be in two places at the same time because it is the most natural piece of matter behaving without the influence of time; it is outside order in nature, it is not directed by anything; the human mind's notions of order and time sequences do not apply to it. You would be shocked by its strange behaviour because you can only judge it with a shallow mind that came to exist after the quantum and is therefore unknown or recognised by it. This is looking deep (speculatively) into matter to the quantum level, so deep that physics cannot include it---but that is how physics itself came to exist and yet it works to the extent of having the capacity to destroy all life on earth.

As the most original matter and the smallest bit of matter that can exist, the quantum is not subject to any of the human concepts about order, time, motion and chemistry as we know them. The quantum existed before the regular cycles we use for time, order, motion, chemistry. This is how Feynman put the same idea: "*The word 'quantum' refers to this peculiar aspect of nature that goes against common sense*"---exactly.[76] It belongs to a universe before the common sense came to exist and is therefore not subject to any of its notions. Common sense refers to common objects of the perceptible universe. The quantum is not part of this universe. It's the most original and basic matter whose many and varied interactions have created phenomena as we learn from QED; it therefore does not know how to behave to suit us as we are part of that phenomena to which it does not belong.

---

[76] See the Introduction to his famous book QED, Princeton, 1985.

We can construct time out of the phenomena in our experience, that's our peculiar luck or curse. It all depends on how you look at it.[77] All the elements for this act of construction are there everywhere in the universe but not as time (to the universe). They are rather events occurring haphazardly under a variety of forces: gravity, space, inertia, motion, heat, chemistry, without conscious control. The religions are right about one thing: the process of human creation requires intelligence. Where did it come from? They claim to know that, but cannot prove how they know it. References to the so-called scriptures are what annoy scientists most.

Again, our instincts expect the quantum to respect time or behave according to time sequences---but the universe has no time. The parameters we use to reckon time are purely accidental events. We know they occur, but cannot think of how anybody could have organised them in the manner we arrange things on earth. As we have come to realise, there seems to be no direction in the universe, no purpose and no logical sequences. Existence is existence; it's just there. Life came as a chemical accident, but, like everything else, it just happened for no purpose at all; and while it seems to us to have been long in existence, in the cosmos at large our period of existence is just a flash, and the earth itself just an infinitesimal dot, not worth bothering about. We see nobody there to worry about it either, unless and until we set things and events in earth-time. Ah, but the universe has no time! Earth time is just that, namely a time system we have created for ourselves on this planet and applicable to this planet alone. That's the conundrum, and I for one find it enormously interesting just pondering it, usually alone, as my religion. Some people weep over life's problems. My advice is to try and find them interesting as events occurring without cosmic

---

[77] To some people life is a blessing; to others it is a curse and regret that nothing in life is worth living for because it won't last and may even end in tears---love for instance. Nobody can enjoys life from birth to death without regrets.

control or significance, and yet so vast and complex that pondering them is itself rewarding as an intoxicating spiritual solace. It's not true that it will make you mad---it'll rather cure your madness!

Nobody is there to infect you with the germs of madness. In nature things happen haphazardly. There are some limited logical sequences like something cold fleeing something hot or ice melting at certain temperatures, but no streams of logical sequences (such as we human beings can construct out of this huge and complex admixture of accidental events), by means of the human mind including the consideration of time---the most essential thing besides life.[78] So it appears that outside a human head time does not exist. According to Russell we 'construct' it ourselves. I agree with Russell absolutely. I do not believe that time does not exist on earth because we are using time everyday; but it certainly cannot be defined logically. For instance, how long is one year defined as a unit of time? When one year passes you know that you've aged one more year but how long is that? Try as we may, we can never define the year on its own logically without using any of its fractions (say, the months or days, which is logically unacceptable).

It's obvious that time does exist on earth; the problem is how to define it. So serious is this problem that we have come to the conclusion that its logical definition amounts to just how it is passing by. So we think we can only know how it is passing by and never what it is. But in the universe at large, although we can spot some of the parameters we use for time on earth (everywhere), no one is constructing time sequences out there through the use of these elements, not from our point of view anyway. All notions of time are carried from earth to apply to the cosmos in breach of the Einstein theory of frames.

---

[78] Whoever solves the problem of time will come closest to knowing the nature of human life of which the mind is the most mysterious, for it's the mind that creates time.

CHAPTER FIVE

TIME AND CONDITIONING IN THE UNIVERSE

Still speculating about time, I think if it is true that every inertial body has to have its own time, then there simply is no time in any part of the universe until you have created your own time and conditioned your mind to its nature, and Einstein made it clear that we can only do so in inertial bodies, not in general relativity. The elements we need to construct our time are not there; and if they are not even available in general relativity then the other world the quantum came from would not have them either. As noted, Einstein divided the cosmos into two[79]: one is where you can have time by sustained regularities (or 'constructed' logical sequences lasting long enough for the human mind to use for its creations), and the other is where, because of the strong gravity, you cannot even see anything anywhere at all to have the necessary regularities

---

[79] Perhaps it should be three: Inertial Frames, General Relativity and the strange world of sub-atomic matter. Similarly, the Two Postulates of special relativity should be three with the addition of time---the parameters that can be used to construct time sequences should be there.

of motion to use for time.[80] For time means from when to when, from one point to another---the year, for instance. Part of the problem in physics come from the improper understanding of Einstein's ideas, for all time is based on regular or repetitive cycles---the year, for instance, and since it is the cycles we count as the rate of the passage of time (like the years) we can never know the true nature of time only how it is passing by. Through our mathematical ingenuity, we've learnt to use cycles to provide units of time. I call this the quantification of time. These are what we use as time: years, hours, minutes and so forth. They merely indicate how time is passing by, obviating the need for complicated theories about how time passes through nature. For a start, our time is not even passing by, or passing through nature. It is discrete and proceeds unit by unit---year by year, minute by minute and so on. But in my experience it seems mankind doesn't want to know about discrete time. It's so enthralled with time passing through nature to the Day of Judgement, and start all over again due to the transmigration of souls. It seems nobody wants to die if he'd not come back to life after his holiday up there! Or live in another world up there. Thus it's easy to get some religious people to commit acts of terrorism in suicidal attacks.

Let me emphasise that what we call time is only how it is passing by---the years for instance, pared down to the seconds. But Einstein is so misunderstood that many writers insist that we need the Minkowski formula to understand relativity; yet the special theory of relativity that concerns us most on this planet had nothing to do with the general theory of relativity and the Einstein use of the four-dimensional continuum of Minkowski in the equations of general relativity. Rather it merely amounts to incorporating time into space to form one entity so as to dispense with the 3+1

---

[80] This logical interpretation was begun by Russell, Eddington and Whitehead. Yours truly is standing on their illustrious shoulders!

formula. There are suggestions about the usefulness of this procedure, but originally it formed no part of the theory of special relativity. In other words, the Einstein theory was complete without Minkowski in so far as special relativity is concerned. The Minkowski theory did not improve special relativity---it's already complete and critically acclaimed. All suggestions to the contrary is evidence of ignorance; for it is obvious to me that only those working closely with relativity understand it without using the Minkowski formula for equating space to time in one equation, yet it was never successful. Professor Eddington was right when he struggled to recall the name of the putative third professor who understood the theory in the initial stages.

Presently as I see it, the religious people want to resurrect the bogey of past, present and future to prove Einstein wrong. They claim that the syndrome causes the flow of the story of history; that history is what the study of the past to the present tells us. Yet past, present and future can be perfectly logically explained as uneducated fiction[81], so Einstein was right: the past is obviously memory; the present is now, carrying the past as historical baggage with it (you never leave your problems or wealth behind you, do you?); and the future is mere speculation. History is the march of these events not time; and the events are still marching on as the continuing story of life. Thus the past is not still existing anywhere to be revisited---it is here with us as the consequences of what happened in the past! History is not seen as time running through nature from the past to the present and so forth. Our time, as the year shows, is discrete. Discrete time cannot march throughout history (or the cosmos) as people like to believe.. Only events have antecedents and consequences. The times are added as the times of occurrence. It is the events that mater. Many of the religious-based mysteries of time can also be resolved. I have published ten

---

[81] Or the illiterate man's understanding of history.

monographs about post-relativity time explaining all these issues, but nobody is showing any interest as people continue to chase their emotional thrills from worthless books and gadgets. I fear that true culture is dying slowly due to the aggressive onslaught of the electronic strangulation

If time is not permeating the cosmos and moving from the past to the present and going on to the future then there is no quandary. You can challenge this theory of time, but by the yearly cycle we should know that time is discrete, from year to year, repeated over and over again for all the centuries; there is only one year in all the universe, repeated to carry on as years; and every unit of time, too, is derived from the year together with its astronomical features. If time is like a thread passing through nature, then it is reasonable to search for theories to account for how it is passing. But we have a time system that is repeated to continue. The year is only one; to have two years we repeat it; to have a thousand we go round the sun a thousand times. Surely everybody can understand how this time passes by as something in procession---units of time following each other. There's no need for a theory to explain how time passes by. This is all we call time. Every unit of time is a fraction of the year as divided with points or astronomical features. Time units have no independent existence; they exist only as fractions of the year no matter how they are derived. The mathematicians have done a good job about this; but there is no mystery; it's plain common sense.

We've nothing else for the reckoning of time except mythologies or counting the days as the passage of time in a primitive manner without theories. I must repeat that there are no years at all in nature existing as something we can just pluck out of the sky and apply to events. Let me repeat again that there is only one year and all other units of time are fractions of the year. To have more years we simply go round the sun again and again; and that is what we know as the passage of time, or it constitutes the passage of time---namely the units of time in procession. This idea solves at once the fearful concept of the passage of time.

We hear so much from writers about the passage of time, but nobody has ever been able to define time. You have to define something before knowing how it behaves. Once time is defined as relation between points, like the year (and that is not in dispute because that is how we get the year, and the year is time), it becomes something proceeding unit by unit or intervals of time in procession causing the continuity of time---like the year increasing in numbers all the way to the centuries.[82] After all, human notions of time come from the year and daily revolutions of the earth, or the day and night system. This time can only proceed unit by unit; so the passage of time is seen as the procession of time units---the year for instance. And as these units are all passing it means all we can ever know of time is how it is passing by and theories of the passage of time are redundant, even humbug when it's incorporated in religious sermons and the Day of Judgement mythology.

Even then (strictly speaking) going round the sun is not time. We use it to show how much time is passing and never the true nature of time; for going round the sun is a physical activity, yet we count them as years because we have nothing else to indicate how much time is passing. The years replicate to become centuries; or they increase in numbers to pass by. *Using the yearly cycle to know how time is passing is not time that is passing. Time does not pass, only the units do; but the units are mere physical cycles.* "A time system", as Professor A.N. Whitehead has said, "is a sequence of

---

[82] The really original thought about the Einstein theory of time is the Lorentz discovery that time varies from place to place. After that the inferences are straightforward---for that is precisely how we get the year---and if time varies then it is not general, fixed or absolute, covering the whole universe and the same everywhere. It is an interesting example of correct definitions leading to permanent solutions of intricate problems. Even the Time Dilation idea it was based on was not correct, yet it helped!

non-interacting moments"----year after year after year, or as pared down to the seconds and the other fractions of the year, and he made this observation in his book entitled The Principle of Relativity.

Everybody on earth agrees that time is mysterious, yet the scientific study of time after Einstein is becoming logically consistent and delectable; the human mind craves logical thought; and if you reason logically from the premise that there is no longer a universal or cosmic time, and that the basic unit of our time is not even definable, so we simply do not know how long the year is and therefore all measures of cosmic time are flawed,[83] time becomes easy to understand.

But try telling this to the cosmologists wasting the taxpayer's money on their pet projects---like smashing atoms. They say it brings technological, economic and medical spin-off---yes, but researchers could achieve the same results over time without wasting billions. How many billions did Bill Gates spend before hitting on his ideas for lucrative ventures?

Brains are what we need. In my opinion the CERN is a complete waste of money. And they're not even sure of their theory. They announced recently that if c has been breached then they might have to re-examine the concept of 4-D geometry to see if it is really true. The point is that the Minkowski ict equation is flawed for being based on the imaginary time coordinate, i, not because of the status of c.

---

[83] Time dilation, the twin paradox, clock paradox, and so forth---none of these is even half as important as showing that time is neither fixed nor absolute but changeable, thus, at once removing it from the realms of religious thought to us down here on the ground as a secular entity. Tracing what it is is proving difficult, but at least we know that it originated from this earth and limited to it. We know how it all began; we can trace it back to when we're mere apes and had no sense of time.

Being a grumpy and infirm 75-year-old great grandfather, I am too old to fear of what could happen to my career; others are probably silent because they have good reason to fear the powers that be! Scientists chasing research funds are as ruthless as the Mafia, probably more so. But there is a lesson here. Science is different from any other calling. It is so open that any theory in any branch of science that is not true cannot be hidden for long.

In case there is any doubt about the Einstein theory of time which states that "There are as many times as there are inertial frames", simply because the parameters used for 'constructing' time are different from one place to another, let me repeat what Professor Eddington said about the matter as a reminder: "Prior to Einstein's researches no doubt was entertained that there existed a 'true even-flowing time' which was unique and universal...Those who still insist on the existence of a unique 'true time' generally rely on the possibility that the resources of experiment are not yet exhausted and that someday a discriminating test may be found. But the off-chance that a future generation may discover a significance in our utterances is scarcely an excuse for making meaningless noises." (Mathematical Theory of Relativity, Ch.1.1.) I repeat this on purpose.

Scientific mysticism has always been part of the problem of time's definition, but now, thanks to Eddington, everybody can be sure that there is no such monster called 'Time Zero' from whence time is supposed to have began and running all through the universe ever since from the past to the present and the future till God calls a halt to the whole damn thing on the day of judgement; a childish fable forged on us as the true and most profound philosophy of existence. To my mind this is intellectually shameful. Religious believers may be gullible, but the rest of us are not that childish to believe an infantile fable like that.

By the same token, 'curved space-time' by which time travel is said to be 'a scientific possibility' by the very people awarding the Eddington Gold Medal to their clever fellows is totally untrue. But sure, physics has got to put its house in order. And the physicists must start with no illusions about our time having any influence in the universe, for the whole earth is only a tiny, tiny, tiny little infinitesimal dot in the milky way, let alone the entire universe; and the psychological-time constructed by the 'worms of the dust' crawling on its surface (as the poets describe us), can hardly influence the universe at large, although I know that greedy publishers wanting to cash in about time travel based on human gullibility, will continue to publish such books (as mentioned above), implying that we and our time can have some kind of influence on the cosmos as a whole. There ought to be a law against the spread of such falsehoods that make people less not more rational. The universe is certainly mysterious; but time is not so strange any more, since we know that at some stage in our lives we simple had no time because we lived like apes on trees. Since we came down we have tried many things for telling the time, the most rational of which is the earth year, pared down to the seconds and the atomic pulses---that's the most logical explanation of time possible and we owe it to Albert Einstein alone. I will now try to sketch what time was before Einstein and what it became after him

CHAPTER SIX

THE NATURE OF TIME BEFORE AND AFTER EINSTEIN

Before Einstein time was supposed to be general and absolute such that any unit of time here is the same everywhere else, ultimately attributed to God. After Einstein we see it as a secular entity that is limited to a frame and discrete because we can reckon its passing only with repetitive cycles. These cycles can only give us discrete units of time, year after year after year and so forth. We pare the year down to fractions ending in the seconds and the atomic oscillations based on the second. As some writers have observed, Einstein's theory of time arose from experimental results and therefore not open to doubt. Now let us look at time before and after the great man in a little detail.

SECTION 6 (A) - TIME BEFORE EINSTEIN

Generally speaking, few writers have studied time seriously with a view to suggesting cogent theories about its nature before Einstein.[84] Even those who made such attempts, like Henry

Bergson, were guilty of assuming that it is just there.[85] That we find it in existence, and that is that. Even the careful logical thinkers fared no better. They made it look synonymous with motion or 'Being', eventually calling it the "irreversible passage of existence". Yet existence is not one; it is multitudinous and individual. And every individual is uniquely separate with his or her own perspectives---no two persons, as Einstein showed with his analysis of simultaneity, perceive one event identically---space and time coordinates are involved. The multitudes of people perceive the world differently. Above all, existence is not altogether passing in tandem. In quantum physics directions are not even fixed; what may be irreversible to you could be the opposite to somebody else looking at the save event from another angle. It is necessary to mention that Einstein also failed to decide how time is created, and stressed only that it is neither fixed nor absolute, adding the most revolutionary idea that it originated from this inertial frame and that there could be as many times as there are inertial frames, thus laying the foundation of secular time.

To digress a little here (taking liberties, the reader might say!) to discuss the thinkers I commend, I have to stress that, contrary to the opinions of some scientists and particularly the pure mathematicians, philosophy is not as time- wasting as is generally

---

[84] Kant also tried. No doubt his numerous followers who believe he's invincible would be looking for his name. He tried his hands at everything in philosophy, and that is the reason he achieved very little of lasting value. Kant always lumped several topics together in sweeping statements, linking physics, physical reality, astronomy, cosmology, logic, linguistics, metaphysics and even psychology in a mesh of contradictory doctrines. As always, Bertrand Russell put it best in his History of Western Philosophy: "To explain Kant's theory of space and time clearly is not easy, because the theory itself is not clear"---I would say the same thing about everything he wrote.

[85] Henri Bergson was better than Kant on the subject of time; a great French philosopher in his day, he's one of the very few brave thinkers to write a book about time, though he spoiled it with thoughts about 'Freewill'---the two don't mix very well. However he regarded space and time as completely independent of each other. They still are.

put about; it is so serious that it shares with theoretical physics the ultimate attempt to formulate credible theories for our understanding of the nature of the external world, or the cosmos as a whole, including psychology and cosmology, or the mind and life ---what happens in our heads and the entire universe of sentient beings. There have been great and valuable contributions to human welfare and material progress from philosophers. Apart from what one may describe as 'the scientific thinkers' like Pascal, Archimedes, Aristotle, Pythagoras and the rest, only a few philosophers have actually made valuable contributions to human progress, but they are there. I am thinking of writers like Plato (with all his faults he discovered the idea that we can never perceive reality as it really is by his most original, first-class supposition called 'The Simile of the Cave'. He also promoted philosophy and taught Aristotle.) Rene Descartes' merits included the Coordinate Geometry, upon which relativity is built, and of course, The Cogito. Our own Bertrand Russell's merits include the new logical theory of 'Denoting'. Henri Bergson speculated that space and time are separate entities. If we had listened to him we would not have wasted so much time and energy over the Minkowski fiction that they are one entity---which I believe is still causing distortions in physics and relativity. Professor A.N. Whitehead and Bertrand Russell also discovered that the world of sense is 'a construction not an inference', which liberated scientific thought to an enormous degree and still bearing fruit. For example, the photons 'construct' images; this idea can be seen as much more rational than supposing that we perceive pre-existing images created by God, which we are even only able to see by the grace of God to be invoked by the mind out of thin air as Plato proposed in his theory of Ideas, and which has been used to justify the existence of God by some of the so-called great religions. My objection is that religion causes wars and sectarian conflicts and therefore is not good for mankind in any way whatsoever. In monumental efforts,

Russell and Whitehead again tried to derive all mathematics and arithmetic from logical premises or axioms; and even though it's judge to be unsuccessful, many logicians believe that it inspired Kurt Gödel's Incompleteness Theorem, otherwise known as The Theorem or The Proof. However, as we have seen, Godel later spoiled his contribution with suggestions that time travel (a notorious religious myth) may be 'scientifically possible' and that even Einstein agreed with him!

The Cartesian Dualism and his notion of God as the Absolute Perfect Being are not mentioned because they are religious notions and religion causes wars and does not advance philosophical knowledge. All the religions borrow from two basic ideas. Due to the fear of death as the end of human life, they borrow from the Pythagorean philosophy of 'The Transmigration of Souls'; and also from the Rene Descartes dichotomy between the soul and the body of man, or Cartesian Dualism as it's called. We don't want any of that because it promotes religion and religion causes wars. Human life is bedevil by a fantastic paradox: the majority of mankind believe that they need to worship and couldn't live without it, quite apart from the hope that it could make death just a transition to a better life; yet the organisations they establish for this worship cause wars and sectarian conflicts that make peaceful life on this planet pretty hazardous.

Returning to our main subject, time, as I have said, had many different meanings for different thinkers before Einstein. I identify four such meanings all of which, due to the importance of time, have become the focus of mass consensus, following, or even civilizations.

First, we have the clever and profound, world-shaking, intellectually brilliant interpretation of time proclaimed by the Irish Prelate, James Ussher that God created the universe at exactly noon on AD 4004, implying that time was already running from the so-called Time Zero, marching on to give us the story of history. For generally history was only understood as the march of time. Let us call this religious view 'Act of Creation'. Even school boys can see that it has not much intelligence to commend it. Yet the majority of mankind, numbering more than seven billion, seriously believe that this is how the world came to be in existence, and worship anything they believe to have 'Created' it. A religious interpretation of time that nevertheless links it close to life as all the logical analysis suggests. There can never be a final, definitive theory of time satisfactory to all mankind. Properly, we should all live with what we believe and forget about the rest---except that time is a powerful agent of causes in science.

Next, Isaac Newton. Newton believed in absolute space and absolute time, and since he was very great in science, everybody able to think followed the great man's definition of time, and it ruled scientific thought until Einstein. Even still now many scientists speak about time as if they think it's running all through the universe like a thread—or that it's general. To be honest, I really consider even those who claim against the facts that time does not exist to be more intellectually respectable than their opponents, mostly scientists and philosophers, who insist that it just is, whatever they mean by that cryptic assertion. In the words of one book reviewer already mentioned above, "Time does not flow, it just is", a biblical language for expressing a biblical myth. For if that is so then what is the use of the earth-year---why should we try so hard to find a logical explanation that ends in the existence of the year? I rather accept Bertrand Russell's theory that it is constructed as "relation between points". Thus sentience is require, because somebody must be there to set the points for the yearly cycle; which means we are sent back to the beginning as the religious people claim that only God or a Supreme Being with infinite powers could guarantee that somebody with the necessary intelligence will be there to do that. So, in the end, we have to realise that nothing in life is easy to explain without an ultimate attribution to a deity, least of all time!

The third movement is the Lorentz/Einstein denial of absolute time and absolute space, demonstrated (or proved) with scientific experiments. Soon after that we got the fourth interpretation of time as his own mathematical interpretation of the Einstein notion of time, called the Minkowski formula for space-time continuum, or 4-D geometry for short. Its brevity belies its momentous effects on scientists. Because it was supposed to be scientific, or the mathematical interpretation of the Lorentz/Einstein experiments, practically all scientists refer to every time and every space as 'space-time', meaning that space has been equated to time---that the two entities have become one. And yet Minkowski could not even define time. Also the very great scientist and mathematician who confirmed the general theory of relativity, as we have noted, Professor Sir Arthur Eddington, the founder of astrophysics, described the theory as arbitrary and fictitious. I have therefore rejected the Minkowski formula as logically untenable, since it is based on imaginary time coordinates. The 4004 Creation of the universe is too dumb to think about, and Newtonian absolute time is abolished by Einstein. My rejection of the Minkowski theory is not mathematical in the sense that it is not written in mathematical symbols but in words. However it is somewhat mathematical in the sense that it is a logical objection and all mathematical statements have to have a logical premise—and it is his premise I am challenging. I am simply pointing out that in the absence of a universal time Minkowski had to show where his imaginary time was coming from; never mind that it's imaginary, but from where? Also how can anything imaginary be relevant in discussions about time? I don't know how to write these thoughts in mathematics so I use ordinary words that everybody can understand.

So we are left with the rational consideration of time Einstein proposed, namely that time is derived from your local space---points together with the ability to count and sentience are required, exactly the way we obtain the year, which is the basic time unit on this planet. For the philosopher, the problem is not that time does not exist since we are using it daily in all sorts of activities, but how we get it, how we 'construct' our time, as Russell put it, because cosmic time is abolished by Einstein. This is the current logical situation about time overall---that is, scientific, philosophical and practical. Now let us look at the specifics, or how some writers have considered the matter.

With regard to the view that time does not exist at all, my understanding of the thesis of Professor Yourgrau's book mentioned above is that the legacy of Godel and Einstein to the effect that time cannot exist under relativity has been scandalously neglected or forgotten by the world. Additionally he says, when that is given its proper due, time travel becomes a scientific possibility.[86] But this cannot be true because the very idea that there can be such a thing as 'The Logic of Time', as proposed here, is derived from Einstein's researches; this logic makes time necessarily discrete, so that we count individual years to get the centuries and our own numerical ages. The fractions of the year too (the seconds, hours, minutes and days) are also separate and individual units of time. Obviously, there is no chain in time; every unit is uniquely separate. That is what is meant by discrete time. However, discrete time, such as we have here on earth, makes suggestions of time travel laughable. Yet the sad fact is that most of

---

[86] It is difficult to see how time travel could be possible if time does not exist. It is also not an honourable intellectual stance to argue that something human beings cannot even avoid does not exist. It may be intellectually difficult to define but that we know something we call 'time' is indisputable. But then we clearly have to understand that intellectuals have a bounding duty to explain the reasons why and how an indefinable entity is in daily use. That is the situation with time, God and life. Of these three basic conundrums time alone seems (not only in existence) but positively decipherable.

the recent writers on time seem to be only interested in theories that make time travel appear to be, as Professor Yourgrau put it, "a scientific possibility". Or to quote him in full: "Godel, the union of Einstein and Kafka, had for the first time in human history proved, from the equations of relativity, that time travel was not a philosopher's fantasy but a scientific possibility." Kurt Godel bears most of the blame. He asserted that his discussions with Einstein had convinced him that time travel is feasible. I concede that scientifically that could be so if time is naturally equated to space; but so long as they are separate entities they cannot 'curve' together as to suggest that time travel could be possible. The Minkowski attempt to merge them with mathematics was not successful---yet even he himself admitted that before that they're separate entities. And yet, as discussed below, in nature we human beings can never make two phenomena one or one entity two.

Thus I wish to assure the reader that there are absolutely no such (logically valid) equations in relativity that could even remotely make time travel seem plausible. In special relativity, there is none at all, and yet that is what concerns us most, since there is nowhere anywhere in general relativity to be capable of giving time there to anybody. The so-called interpreters of general relativity carry earth time there in clear breach of the Einstein theory of frames. What happened is that mathematicians coerced Einstein to incorporate the Minkowski formula for 4-D geometry into the field equations of general relativity---yet the Minkowski formula is logically flawed and therefore completely unacceptable as the basis for altering physical reality to one of four-dimensional space, in fact, it bears no relations at all to physical reality and looks beautiful only in his own mathematics, which is considered arbitrary.

Let me point out that what the mathematicians wanted was a new formula to replace the 3+1 system so that, as they do now (even though they know that it is not logically valid), they could write one equation to represent time, space and matter as 's=ct...' It was an extraordinary demand to address to inanimate nature, and would be funny, something like a silly prank, if it were not so serious.[87] Man is an orphan because he does not know whence he came. He is a clever orphan/ape on a lump of rock; and for such a creature to demand something from nature by way of his own mathematics is nonsense. Yet they went ahead, and Minkowski obliged. However, Professor Yourgrau himself has quoted David Hilbert as saying Einstein did not believe in the concept of 4-D geometry or four-dimensional space. Therefore, space and time remain separate entities, especially on any inertial frame like the earth, as Einstein made them in special relativity.

In all history, time has been regarded as the same as existence, hence the age-old definition of time as 'the irreversible passage of existence'. In effect, it equates motion with time.[88] In addition, since nobody knew (or still have any idea) how life came to be, no one worried about the nature of time which is closely associated with it. Instead time and life were lumped together as constituting the eternal mystery on earth. The genius (unique insight) was the suggestion that it does not even exist until you have invented it out of the parameters in your environment. This is the Einsteinian

---

[87] In all life (from the beginning to the present), we have come to realise that logically and realistically, addressing demands to inanimate matter is a sheer waste of time---like all prayers. It may help through psychology but not through the alterations of material reality. Physics means material movement or manipulation, without that material changes on the prayers of another material entity (man) is sheer nonsense.

[88] I have explained above why motion cannot be equated to time, although it can be used to reckon time either singly (like the earth's motions) or in combination with other factors, such as chemistry. So I see the attempted equation of motion to time as an interesting paradox of contradictions: if motion is equated to time and yet the motion takes place in space how can the time be equated to space? Above all, time can only be had through the application of points to space! We're seeking to bind our thoughts with stultifying contradictions.

notion that Kurt Godel misinterpreted. For time does exist on earth; the fact, however, is that it did not exist before we came down from the trees. It means we created it; and since its creation it has come to dominate our minds and everything we do, like water. Nobody is born demanding water in order to live. But once we taste it we realise that life cannot go on without it.

The mathematicians used astronomy (being the natural features of the world, as perceived) to construct clocks, and that was it. We simply used the time provided by the clock makers. This time was taken as general and absolute; it was assumed to have originated from divine sources; and since a centrally imposed time system could not be different in different places, one second here was supposed to be one second everywhere else.[89] Then Einstein burst on the scene with his new idea of time, showing that the old idea of time was religiously imposed and not really true---obviously it is neither general, fixed nor absolute. He proved this with experiments.

Given the supreme importance of time in human affairs, the irrational view of time (before Einstein), influenced all life and all ideas including religion and historical narratives adversely, spawning mythologies many of which are still with us today with their own consequences, some of which are even detrimental to human existence. One of these is that history is the march of time---marching since the 'Dawn of Time'---instead of recognising that history has been the story of how life has been lived through successive events since the 'Dawn of Existence', or intelligence, or the first acts of sentient beings on earth, from which acts all successive events have flowed as the inevitable consequences thus giving us the continuing story of human history on this planet.

---

[89] Nowadays, I guess no educated person will believe that one second here is the same everywhere. But in the past even Newton accepted the religious view that time is absolute or fixed for the whole universe.

Now, rationally, time is not seen as marching on and taking us with it; rather we think we are marching on event by event and recording them as they occur at certain dates and times.

SECTION 6 (B) - WHAT IS SECULAR TIME?

First of all, (before we discuss time after Einstein proper), we have to define what is meant by secular time. The Einstein theory of time is called 'secular' in the sense that it is traced or deduced consistently from premises based on material reality without any attribution to any god or deities. Thus what has to be explained is the phrase 'secular time', and that is this: everything we refer to as time must be a recognised unit of time. The word time has no meaning without quantification---see the Appendix 'Time and Quantified Time' below. People usually mention the word time to mean the passage of anything---events, moments, even sitting still---but that is not logical thought or attempts to define time as it is used in all activities in society or the world. Time in science, logic and life (or what of it that we need to understand or explain in ordinary linguistic usage), is presumed to be either materially based or mathematically calculated from features of the earth logically, as opposed to time that is simply imagined or mentioned as the term for any passing moments. One common example is time in dreams; another is just referring to time in ordinary conversations that are generally understood to mean any moment of passing consciousness---the shortest time, the longest time, and so forth.

In society or real life (as opposed to these vague instances of mentioning time without definition), it is necessary that everything we refer to as time---every unit of time in use---is logically traceable and derived from the periods of the revolutions of the earth and its long orbits of the sun---usually shortened to the phrase 'motions of the earth'. Not long ago the individual units of time, the hours and minutes and seconds, were regarded as divine; God had actually created them as independent entities.[90] They're easily explained with mathematics as fractions of the year that would not exist without it, yet the mischief-makers claimed they had independent existence. Even still now many scientists continue to speak of time units as if they're mysterious. But of course they're not. The year is determinate. All the units of time are formulated as fractions of the year and its astronomical features so that a certain number of each unit of time will add up to exactly one year to coincide with the complete orbit of the earth round the sun. Thus there are no thirteen months; no 54 weeks; no 368 days in one year. The units do not go on after the year; they are all recounted from the base of one at the end of the year.

The 24-hour periods and the long orbit of the sun provide considerable periods of planning time for all activities: time to catch a Bus, plane or train; time to eat; time to walk a distance; time for work; time for sports---time for doing anything at all. All of these are derived from either the 24-hour periods or the earth-year. In secular time we realised that a mathematical or logical explanation was required, and our own Bertrand Russell, as the world's most recent great philosopher (who was also a logician, writer of genius and great mathematician), provided the world with an appropriate theory called 'relation between points'; that's the only time that can be programmed into the clock, and we all know that time in the clock is the only reliable time.

Time, he said, is a construction.[91] Together with his collaborator, Professor A.N. Whitehead, he also interpreted the

---

[90] I concede that coming fresh to the study of time, the phrase 'time units' would require an explanation. But there is one available.

[91] Leibniz also said it's a succession, missing the term 'units'---time units in a succession as the answer to the problem of the passage of time. Nevertheless I mention this to show that Leibniz was also very clever, no wonder he and Newton quarrelled about mathematics; intellectually both were almost on the same level. We're lucky that these men come in doubles: Russell and Einstein, Newton and Leibniz, Plato and Aristotle, etc.

world of sense as 'a construction' rather than an inference, to overcome the old practice of philosophers inferring all things and connections in their minds as their logical definitions of physical reality, contrary to the physical reality discovered in physics, or the actual physical analysis of what we perceive and can also infer from what is perceived. So secular time refers to the time system we can consistently trace from the mathematical, logical and visual premises all combined. Every mention of the Einstein theory of time is to be understood as 'secular time'---traceable from material reality without mythologies. The only system of time we can programme into a clock. It is conceivable that Russell gained his insight by asking the question, if cosmic time is abandoned then what is put in the clock as time? Given sufficient logical acumen, everybody can deduce that the new theory is calculated (and can only be calculated) from the motions and physical features of the planet we live on. That's the meaning of secular time.

SECTION 6 (C) - TIME AFTER EINSTEIN

Albert Einstein changed the debate about time for good with his division of the universe into two distinct categories, governed by different natural laws: (1) General existence, or general relativity, where objects or matter just existed and whirled around under the influence of gravity without any conscious directions; and (2) special existence 'in' special relativity frames or bodies, where the two postulates and time applied, perspectives arise and intelligence and life can flourish in response to the intelligent use of available resources for civilizations to rise and fall---or generally for life to flourish as it cannot do in the whirling flux of general relativity; thus creating the never-ending chain of events known as history or the continuing story of human existence. Since civilizations arise upon definitions of time as mentioned above, the reader can see that only the very rational, scientific civilization can be consistent with the new concept of time.

Einstein did not deliberately set out to change our view of time.[92] It was an accident discovered by Lorentz. He said the Lorentz concept of local time may be regarded as 'time, pure and simple'. His genius made it sound simple, but it was the beginning of the most profound revolution in human thought. It was unique; the nearest idea pointing to the origin and purpose of life because time is the second most important thing in the universe, bar the life itself.[93] And the two are inseparable. No thinker has any idea as to which is which or which of them came first. Personally I believe everything in human experience is generated by the brain in us; this implies that all human creations are secondary to the existence of life part of which is the brain.

Of course, on the other hand, Einstein did deliberately (and even contrary to classical physics), set out to change our views of the universe. The result was the theory of frames with which we are now familiar. It divided the universe into two distinct categories. One is general relativity, where there is nowhere anywhere for life to evolve and flourish; the other is the inertial frame, where life is possible and civilizations can rise and fall. Time is required in this second division of the universe; and the local time idea was just the thing to suit inertial frames. I think we should now write time as the

---

[92] Time is the essence of human life and the cause of the rise of all civilizations, but the irony is that, what we call time is only how it is passing by! Indirectly it is life we're discussing. For time is only one of the agents we employ for the proper organisation of life. There are others: water, heat, electricity, gases, even volcanoes. To define time to the bottom is to find the essence of life—which is physio/chemistry, motion and intelligence all taken together. There is no doubt that the familiar ingredients of evolution were present. The mystery is how they came together so perfectly to result in a sentient being with intelligence to probe nature and improve his life with the materials found on the planet without any help from any quarters. Of course religious leaders claim to have the answer; but it turns out to be just what they imagine in their dreams---the reason it is considered foolish for gullible people to worship any god at all.

[93] If time is physio/chemistry then so is life, as the theory of evolution makes clear because the two are very closely related and inter-connected.

third postulate to add to the two original postulates of the special theory of relativity.

The problem thence is to discover how our own time began, not as a version of a universal time, but a time limited to our frame, a major kind of philosophical inquiry since time is inseparable from life. That old idea was a mistake; yet everybody in science is still considering time as if it is something generally in existence and our time is a version of it.[94] Thus, the Minkowski formula for 4-D geometry is defined as incorporating time in the three dimensions of space to create 'space-time', the merging of space with time the end result of which is to give us what we call time as 'space-time'! In the absence of a universal time, where is the time incorporated into the natural dimensions of space coming from, if the end result is only to create time again? Using time that does not exist naturally or universally to create space-time as time by means of mathematics--what sort of logical reasoning is that?[95] I am silly enough to let it bother me a lot; I really do not believe that it worries anybody else since everything I write is never even read in manuscript let alone published---my long suffering son has had to do it on my behalf, yet he's only an engineer! It may well be that people are literally afraid of time---afraid to offend the Gods. Yet, actually, the nature of our time is easily deduced from elementary logic.

In fact, as Russell put it, "There is no longer a universal time..." Thus, he asked, "What is measured by a clock?" Yet the question is wrong. The clock does not measure time. It rather reproduces units of time specifically programmed into it for

---

[94] Culturally it does not seem likely that we can ever abandon all traditional references to time; but in analysis scholars should try to do better.

[95] Yet this space-time is compounded of the materials found on the earth only for the convenience of the mathematician, according to Bertrand Russell, who, of course, did know a few things about mathematics.

reproduction. That is the reason it works in units only---second, second, second, and so forth. The real problem is how the units of time programmed into the clock are derived in the absence of a universal time. This was answered by Russell himself though he did not realise it. He defined time and space-time as 'relations between points'. He was right. That is how the seconds and years are created. Relation between points that are constructed out of the natural parameters we find on earth, hence the notion that it must be the logic of time in the universe as a whole. We create our own time. While a universal time does not exist, any 'Beings' in the universe would have to invent a time system similar to the one we have invented on earth.

But if the year is abandoned, will time be going on still? In my view, the answer must be both yes and no. Yes, because events and activities like chemistry, ebb and flow, growth and decay will still occur, though they could be seen as mere chemistry, physical and organic. But no, not the sort of time we are accustomed to since we couldn't tell 'how much time?'. We would have no method of telling 'the time'. For we count the yearly cycles and pare down a year to the seconds to be able to tell how much time. In the absence of that system we're lost. Our civilization will die. We'll have no philosophy of nature to live by. That's how important time is---and we created it, not God.

The year, of course, is basic. The seconds and all other units of time are derived from the year with mathematics as fractions thereof. Everything depends on the use of points. We use points to get the year. The fact that it is repeated over and over again to give us all the centuries means our time is determinate---in other words, our time is discrete. The essence of a discrete time is ended when the units are expended, thus we have to repeat the yearly cycle for our time to continue all the way to the centuries by replication. In addition, the system runs all through our units of time: from seconds to the minute, minutes to hours, hours to days and so forth. Our time is not a thread running through nature as we used to think; what we have found through experiments is that it consists of a chain of individual units created with points or mathematics in association with astronomy and the essential features of the globe; as such it consists of separate moments, as Professor A.N. Whitehead has stated in his book, "The Principle of Relativity.

It also means time is not known ahead; what we call time, say the year, is known after it has passed---e.g. the year is not ended until 31st December. That is when we can have a whole year. Then we have to start another year. The same principle applies to all the other units of time derived from the year, including, as I keep reminding the reader, the atomic units of time, because they have always to be related to the second to make sense. Secondly, time cannot be seen as the cause of events; events are physically caused; the times are recorded as the periods during which they occurred. Thirdly, time created with points and which is not part of a universal time, cannot have anything even remotely to do with what the religious leaders dream up about the nature of time. Fourthly, the passage of a time system produced with points unit by unit, as the year shows, requires no arrow or arrows to pass through nature: the units replicate to pass by---precisely as the years replicate to become centuries. All that remains for time to take its rightful place in science as a rational subject is for mankind to wean itself from the 'sweet' religious suggestions about time (what Professor Eddington called 'even-flowing time'), since the true facts are now well known: we don't know what it is, except to guess that it is the product of sentience, physio/chemistry and motion; but we know how it begins; we also know how it passes by---second, second, second; or year after year after year; and we know how it will end, that is, when our planet ceases to support sentient beings who can count the orbits of the sun as 'years'. Religion has nothing to do with it. The arrows of time for its passage through nature is redundant; and is definitely not universally existing in the cosmos because without knowing how to count the orbits of the sun as years, there could be no years just bland existence.

In conclusion, let me point out that, if the distortion from the Minkowski formula is eradicated, the question of time under relativity becomes simple, exactly as Einstein put it, namely 'pure and simple'. Here are the basic facts: (1) There is no longer a universal time so we have to search for the origins of our time because; (2) every 'body' or inertial frame has got to have its own time ; (3) under relativity the all embracing time is a construction, like the all-embracing space; (4) both the earth-year's time and the atomic time use regular or repetitive motions to track time----that means they can only track passing time since the pulses or motions can be counted 'after' they have occurred and not before. We put all this together and get the notion that time cannot be logically defined, which means that what we call time are units of passing time. They are units because we get them from repetitive motions or cycles---and that is the reason it is passing time, simply because, of course, these regular cycles are passing. One after another (or year after another year), there is nothing more to time. All units of time are derived from the year; even the atomic units are related to the second to make any temporal sense. So every unit of time is a fraction of the year.

Thus, in the end, since the years are our only means of noting the passage of time, the explanation of time was rather easier than going through all those complicated mathematical and physical theories of arrows, mysticism and divinity. It is conceded that time is mysterious. It is even assumed to be the last refuge of God after Charles Darwin, since many people believe that time's deep, fearful and intimidating mystery goes beyond human comprehension.[96] Yet it is rather ironic that we have been using the orbits of the sun for time without realising that it is all of our time---mere physical cycles counted as years, centuries, millennia---because we thought we were measuring our version of time out of general time permeating the cosmos, the provenance of which was assumed to be nothing but divine. Yet once we learn that there is no longer a universal time (thanks to Einstein), and that we do not measure time at all, the orbits of the sun appear in a new light: we count the mere physical orbits as our ultimate units of time (the years), out of which all other units are derived.

---

[96] Time is the closest thing to the nature of life. Whoever solves the mystery of time must know more about life than the rest of us, and that honour goes to Albert Einstein for observing that the Lorentz 'local time' notion can be defined as time, 'pure and simple'. Until then everybody assumed that time originated from the cosmos and probably of divine provenance. If anybody can invent time, or his time, then its origins must be human. And that was a great philosophic insight.

CHAPTER SEVEN

ENTROPY, GRAVITY AND TIME

Concerning entropy as racing with time to our doom, or time causing the increase in entropy to our total extinction, and also the effects of gravity on the whole of time through a single clock[97], I would say this: it all depends on how time is seen or defined, but I believe the idea that entropy and time moving all through history to just one sort of destination which is the death of heat and activity so that energy and life will come to an end is religious and echoes the Day of Judgement mythological sermons.[98] Can this sort of

---

[97] Note that when the matter is stated clearly like this the quandary is partly dissolved!

[98] I regard it as unfair for anybody to use time as it is and define it as it cannot be, or contrary to logic. For we all use time in the clock. This time is discrete---second, second, second; and it is based on the earth-year, which is also discrete. This earth-year is only one, yet we have had several centuries, which means this one year is replicated, to amount to 'years' of any number. The same discrete year is divided into several discrete units. Anybody who uses time cannot deny this, and yet people define time or discuss it as if they know that it is running all through the cosmos like some kind of a thread as preached by the religions. It's unfair for them to use time in the clock. Let them invent their own system of reckoning the sort of time they subscribe to in accordance with their religious beliefs.

thing happen in places like the sun where there are no religious inventors of such theories?

Now that we know time is not universal, any objection to the logically deduced concept of the time we have must be flawed or just religious sentiments not to be taken intellectually seriously. Similarly, I think gravity is given far more scientific kudos than it deserves. As important as it is, if gravity affects any clock then such events should be interpreted like the clock paradox, since it is known that acceleration affects some clocks' performance. For it cannot be stressed too strongly that any clock's performance under any circumstances is not the whole of time per se, but that of the relevant clock alone.

The objection to what we now have as mankind's most logical theory of time (since we learnt that time is neither general nor absolute and that it does not run all through the universe and the same everywhere), should be regarded as either fantasy based on ancient traditions, a religious idea or mythology disguised as rational thought.[99] Einstein is said to have failed to establish the true nature of time because he could not explain past, present and future. Yet he did explain it absolutely clearly. He described it rightly as "stubbornly persistent illusion"; and, as always, he's absolutely right because it is still stubbornly persisting and remains an illusion. So far I think the reader of this little book will have realised that time, as defined here, cannot be entropy marching on to our doom, neither can gravity affect discrete time. <u>The truth is that all the ideas circulating about time refer to a universal time, fixed, general and absolute, covering the whole universe in one</u>

---

[99] The three essential facets of the traditional concept of time as an absolute entity ultimately caused by providential command should always be emphasised. They're that, (a) time is fixed and unchangeable. (b) That it is therefore absolute and permanent. (c) That it is generally permeating the whole universe and the same everywhere---so that a second here is a second everywhere else. Whoever works in science and is not offended by these explanations of time as meaningless fables, must be religiously inclined.

format. Discrete time is as yet unknown outside the highest circles in science and philosophy, and since time is regarded as mysterious the general public does not get near. They even claim it brings a curse when probed too deeply! The problem is that absolute time is abolished under relativity, while discrete or secular time has been proved in Einstein's researches, as Professor Eddington has confirmed. The jury shouldn't have much problem deciding this issue in favour of secular time.

All I want to add is that in our religious and ignorant pre-scientific existence, when the nature of time was a mystery with all thinkers displaying their 'massive' intelligence by producing arguments to deepen the mystery about absolute time rigidly running all through the cosmos and giving us the story of history and so forth (note that the story of history was more the march of time than the events engaging people), it was perhaps intellectually permissible to regard past, present and future as significant topics in the discussion about time. But it's sheer monstrous and offensive impudence to confront a man like Einstein, who had only recently discovered after experiments that absolute time did not exist, to explain past, present and future. He's right to reply that it's an illusion.

All the same, we do have serious problems with time; problems that are leading even good scientists astray so much that they're questioning the Einstein notion of time and preaching that time travel is 'a scientific possibility' while claiming at the same time that Godel has proved (beyond doubt) that under relativity time cannot exist---a contradiction since we have time and also continue to live in a special relativity metric. Really, some scientists are behaving like a bunch of philosophical amateurs. How can anybody travel by something that cannot exist?

Of the numerous concerns about time I have selected only a few topics for discussion. Nobody except a super human being can answer all the queries, legends and myths of time. But Time Dilation, The Clock Paradox, The Twins Paradox, entropy regarded in science as showing the direction of time and the effects of gravity on time appear to me to worry serious scientists so much that I have decided to mention them here and comment on them briefly.

The order of time has already been discussed. Frankly, most of the issues that bother scientists are not really matters of serious concern. What we thank Einstein for is the permanent liberation from the oppressive restriction on the human mind by absolute time running all through the cosmos from a beginning to a predetermined end. It is this mythical fable that spawned the most intricate philosophies of time from Plato to Kant and modern science, simply because there was no way round its iron grip (enslavement) of the human mind. Even when Bergson said, correctly, that space and time were separate entities, it didn't really register much in the way of the rationalisation of space and time before Einstein.[100] If it did, Minkowski would have had to struggle to make any impact, rather than the euphoria with which his theory was greeted to the extent of coercing Einstein to mention it in general relativity---even though we are now told that he did not accept or understand it. One of the best news I have ever heard. Yet, according to Dr. Gribbin (world-famous writer on science), it is still regarded as the best way to understand relativity---if so then since the Minkowski theory is logically flawed, relativity is still not

---

[100] To be honest, Einstein was different, very different from almost all other thinkers past, present and future, in that he worked things out in his own mind into a logically coherent explanation of phenomena before coming out with it. He's almost as philosophical as a scientist, with the added benefit to us ordinary men that he gave proof of his theories to qualify them as science. The person I have described should not surprise anybody as a very unique genius, and if I did not mention his name, many scholars could easily have thought it's him. When he's given the good news that his general relativity had been confirmed he merely casually replied that the theory is right anyway! I think if it had not been confirmed he'd still have insisted that it's right.

properly understood! I know the mathematicians would want to lynch me by the nearest lamp post, but there you are. That's a deliberate choice. Some people die committing crimes; I wouldn't mind dying in defence of Albert Einstein, for death awaits us all anyway.

(a) I believe Time Dilation is what inspired the Einstein theory of frames of which time forms a part. Even Lorentz saw that. He later confessed that he did not discover the special theory of relativity because he failed to attach due importance to his own discovery: "The chief cause of my failure [in discovering special relativity] was my clinging to the idea that only the variable t can be considered as the true time and that my local time t^1 must be regarded as no more than an auxiliary mathematical quantity".[101]"

But what is time dilation? It is called "the dilation of time as a measure of moving clocks". Actually time is never diluted or ever dilated by anything except the earth's movements. From time to time the clocks are adjusted to accord with changes in the earth's movements, but the changes are usually so small that only the scientists involve worry about them. So what happened in the Lorentz experiment was pure chance, and it wasn't the dilation of time at all. Through ignorance that is what it was called. This ignorance and confusion was common before the Einstein theory of time. As Professor Eddington has observed, everybody was wrong about the nature of time. Einstein is not called philosopher/scientist for nothing.[102]

---

[101] Abraham Pais, opp. Cit Ch. 8.

[102] It's generally supposed that the clock records time, or that every clock is always recording time exactly like every other clock---but that cannot be true; elementary knowledge of mechanics makes that clearly impossible. The clock does not spawn time originally such that any clock's time is time. Any clock's time is the time it is programmed to produce. This is easily proved. If a clock is running fast, it will mislead you about when to catch you flight. Now, if any clock is running accurately so as to record the exact number of seconds to coincide with the full orbit of the sun, then it is an

There were two clocks involved. The stationary one worked normally; but when the people outside looked at the moving one they found that it was slow, or seemed to be moving slowly. If anything at all this is 'clock dilation' (one clock seemed to be functioning erratically not the dilation of the whole of the time system on earth. Yet this episode has been paraded about that it shows how time dilates with speed, thus deepening the mysteries about time, and God knows it is already very scary to some people. Mathematics can be used to smooth out the differences caused by the moving clock. However, the best solution is the Einstein theory of frames---the moving clock is in a different frame of nature or the universe.

There is no theory or action (even mathematics) of any kind that can dilate or dilute all time from the mechanical functions of one clock alone, for time is physical as well as psychological. The two must agree in visual displays to amount to recognisable time.

It should be clearly understood that our time is based on the earth cycle of the sun; and also all the units of time including the atomic pulses that are based on the second are fractions of the year; you cannot change any of these units without altering the motions of the earth. Again it all depends on how time is defined. And even if Atlas tried with all his might and failed to change the directions of the earth, I doubt that mathematicians can do so with pencils. Their vanity is greater than the might of a superman, I know; but even they themselves have long realised (if they'd be honest) that

---

accurate clock. But it's never possible to tell from any clock whether it is doing exactly that without fault. Factor in the likely mechanical errors common to all mechanical devices, and it becomes obvious that a clock is programmed to reproduce units of time deliberately put into it for reproduction. If it does it correctly then it's year will coincide exactly with the earth's complete orbit, if not, then it is errant and therefore unreliable. Time dilation was taken as a metaphysical statement about time as a whole---that was a mistake. The irony is that it led Einstein to the discovery of the greatest and most fundamental theory in all history, that time is neither general, fixed nor absolute.

philosophers helped to advance science more than mathematicians whose basic instincts, as Newton said, are to destroy other people's theories. At best what happened in the Lorentz experiment is 'clock dilation' not the whole of time as such. But as clock dilation, we know that a cock, any clock, does not create time; it cannot produce time units out of the sky. A clock is manufactured to reproduce units of time obtained from a breakup of the year and deliberately programmed into the clock with a specific mechanism to reproduce it in specific units, so that a certain number of these units will amount to one year exactly---and start again for another year. This is the reason we countdown to a new year in dramatic seconds. A second is part of a year. For example, one second to midnight on 31st December is this year still; one second after midnight is next year, or part of the coming year, and so we'd have one second less to go for another new year celebrations.

In conclusion, l wish to make some of the complex strands in the debate about time dilation a little bit clearer---as far as I can. Scientists are convinced that it proves that the passage of time varies according to the speed of the observer. This cannot be true because nobody knows the nature of time to make such a categorical assertion about it. Always when a discovery is made in science the mathematical interpreters set to work. In the case of time dilation they have concluded that time slows with speed---spawning the twins paradox, clocks paradox and time travel, all the way back to the revival of the Pythagorean Transmigration of souls, dogmatic religious sermons, Day of Judgement, et al. Mankind is basically stupid. We never get over the fear of death and allows it to rule our minds incessantly.

Now, initially the interpretation of the Lorentz discovery of t^1 (or local time concept) was that: (a) the passage of time varies according to the speed of the observer; and (b) the total effect is that time slows with speed. These were regarded as discoveries rather than some writer's fallible interpretations. Both are wrong. The passage of time as a whole does not vary or slow according to speed one way or the other. This was an observation of the functions of only two clocks. In science and logic you have to allow for the possibility that something was wrong with one or other of the clocks---not the whole of time per se. That is set with the motions of the earth and can only vary if the earth's motions are changed. That is the reason we have to adjust our clocks occasionally when the motions of the earth changed. For that is the only metaphysical change in the nature of time we know. With regard to time dilation, the carriers of the moving clock saw no variations, but observers from outside the moving vehicle did. So Einstein conceived his theory of frames to account for the variation. The errant clock was in a different frame because the universe is basically fragmented. It's one of the lucky moments in human history when an insignificant fact can lead to major discoveries and momentous consequences. Lorentz realised his mistake (and said so later on), that either of the clocks in his experiment could falter; but his local time concept was significant. Einstein said it should be taken as 'time, pure and simple'. It showed that time varies according to space or your locality---most shocking and most revolutionary because time was supposed on the highest religious and Newtonian authority to be fixed, absolute and general, covering the whole universe and the same everywhere. It's so mysterious that this religious interpretation satisfied everybody, even though we derived time from parameters set by ourselves, and also it did not exist before we came down from the trees and grew wiser. Thereafter, the new interpretation of time dilation is that it is not variable according to speed but according to locality or local condition, in short according to space in accordance with the theory of frames; for according to the theory of frames, every space has its own unique parameters for "constructing" its own time system---

applicable only to that frame. Therefore "There are as many times as there are inertial frames".

(b) The clock paradox is almost similar in concoction and interpretations. Einstein pointed out that only one of the clocks had experienced acceleration---so it was again a matter of an individual clock working strangely, not the whole of time going off in a crazy manner. Furthermore, and this is crucial, even if it's one clock going off in a crazy manner, real time would not be affected because clocks do not control time. They merely reproduce units of time obtained elsewhere (from the breakdown of the earth-year and programmed into the clock.) Of course we only note time by the clock; that is so but only when any clock is working accurately---and it takes time and effort to find that out. I would call for a little kindness to be extended to the old thinkers locked in the stiff embrace of absolute time in that the new definition of time was not known to them.

(c) The twins paradox. The mathematicians are very pleased with this one; it holds out the possibility that time travel is really 'a scientific possibility', some would say it's already an achievable thing waiting for an adventurous tycoon (another Richard Branson) to demonstrate it out there for all to see.[103] But, as we all know, pure mathematicians are prone to this kind of fantasy---it's a crazy job, only mad or half mad people do it. Otherwise the twin paradox is not worthy of serious discussion and should be taken as just another of those stubborn, persistent illusions. If the clock paradox and Time dilations are correctly dismissed as aberrations then the twins paradox based on them should not be listed at all as one of the mysteries of time that scientists are seeking answers to.[104] The

---

[103] I would bet that it won't be there for all to see, because when he sets out that'd be the last anybody would see of him!

[104] I respect their views and appreciate their concerns; I have written several books to try and answer some of them. But, as I keep reminding them, in an experimental science we

world of learning consists of two separate strands: one is the speculative and the other is the scientific out of which we get our technology as substantially true of the external world. In the speculative branch of learning people are free to think the unthinkable. It is the duty of science to prove them right or wrong---but the speculative is as necessary as the scientific, for it's the only way man can have knowledge, so long as you learn to tell the difference. Unfortunately not many people can do that successfully.

The mathematicians promote the twins paradox as some kind of their own comradely pastime which delights them enormously; unfortunately it causes distortions in philosophy and theoretical physics. They argue that everybody has time in him because there is space in everybody and space and time constitute one entity as proposed by Minkowski. This however is not true. The Minkowski theory is logically untenably. Secondly, although it seems to some people that the body reacts to time, but in reality there can physically be no mechanism in the body ticking time away, simple because any such mechanism would not know what time to copy since there are no standard time units. Many people still believe that there are such units of time as part of the belief in the existence of God, but they're wrong in logic and I hope this little book's arguments will be helpful to them.

Mechanically every clock works alone; it's not wired to work in tandem with all clocks; so an internal clock would not be able to keep up with any specific kind of time; otherwise the question is "whose time?" or "which time?" it would be copying? There is no longer a universal time that is the same everywhere. When biologists talk of internal time they are referring to the tendency of the organs to react to stimuli in a variety of periods in accordance with the age and gender of the person. For instance, a child heals

---

have got to discover properties not assign them from our livery imaginations.

quicker (meaning in a shorter time) than an adult, and an adult far more so than an old person.

In spite of all that, the mystery-makers called mathematicians claim that we have internal clocks; these are linked to space as Minkowski suggested, so they have effects on the ageing process. They then jump to the other, more famous, fallacy, namely, that time slows down with speed, and conclude that as time dilation and the clocks paradox slow our internal clocks down we are bound to age slowly---hence the twins paradox. But physics has not progressed to the extent of sending men to the moon and back and become so fearful with the bomb and other weapons by propagating mythologies simply because they suit the fancy of mystery-making mathematicians working as closet magicians. Science loses its logical purity when ideas are promoted for their mathematical beauty rather than objective truths. To put it as nicely as possible, human beings were existing long before time was invented for the clock; and we've not yet gone through the necessary surgery to implant clocks in our guts.

(1) The ageing process is not wired to move in tandem with a time system that is known to be based on the motions of the earth so that, as far as the ageing process hidden in mankind is concerned, this time is external and even unknown. The phrase 'internal clock' does not refer to any particular system of time physically planted in our guts, but merely refers to the fact that our internal processes can be set to external time to indicate how long they last, or how often they occur---very useful in medicine not metaphysics. The famous Circadian Rhythm refers to how the body adjusts its activities (rhythms) to the world's alternating day and night system, or how the body's functions are affected by the day and night revolutions of the earth; and we can all agree that they're bound to be affected or influenced by the earth's revolutions. After

all the whole idea of having a logical time system is based on the Day and Night system.

(2) There is no longer a universal time that is the same everywhere, so there are no standard units of time in people wired to accord with external time and the mechanics of space travel. But, obviously, if one lives in the tropics he or she has a different biological system, even (dare I say it?) a different skin colour. What happens to the outside of our bodies are reflected inside us; so the day and night system affects how our bodies function, and since our time is based on this system, some activities follow the syndrome and appear to be time-controlled. A single paragraph in a good book on biology will make this clear. The phrase 'Body clock' refers to how the body regulates its activities to accord with the regular Day&Night system. Matters concerning clocks are not questions disputed in philosophy, so long as they are not labelled as 'time'. They will show our system of time (demonstration of time) but do not constitute the time. The time is obtained elsewhere; the clocks are programmed to produce certain units of time to accord with the orbit of the earth round the sun. The clocks may do this mechanically accurately or not; but their errors and malfunctions should not be attributed to the entire time system---hence the clocks paradox does not affect all time. Biological and body clocks are proper subjects for scientific study: the body clock shows how the body reacts to the Day&Night system; the biological clock shows how the different organs in the body regulate their functions by their own rhythms, such that the liver and kidneys have different periods for reacting to poisons, and so on. As I have said, little by little many of the mysteries of time are being deciphered in science and philosophy; and even those science and logical thought cannot reach give us no grounds to suppose that time is in any way divine. It does not mean anybody is claiming that time is no longer mysterious; it is, but so are the function of the liver, the kidney, the inner ear, the eyes and so forth. Nobody can explain why they have

to be there except that they form part of the body to enable it to function properly---yet to what effect since we'll all eventually die? All we demand is that nobody should go about preaching unscientific sermons about anything in nature and impose dangerous rituals on human beings, because nobody knows anything for sure. As far as time is concerned, like many things in life, we have eventually got a fair idea of time's origins and nature through scientific researches. That is the reason for Eddington's severe condemnation of those who still think time is passing through nature like a thread.

(3) By the same token, (since there is no standard units of time) it'd be impossible for a clock inside us to accord with clocks outside the guts of man. In any case, the Minkowski theory upon which this theory is based is described as arbitrary and fictitious. Science is impersonal; we have no means of altering its conclusions; but there are adequate methods for neutralising or mitigating most of its ill effects. They do not include falsifying results to make them (religiously) palatable. That won't move a tram, let alone the rockets to transport man around the solar system.

ENTROPY AND TIME 8

The theory is that increase in entropy leads to the dearth of heat and activity and that this will eventually result in our extinction; and yet as time goes on entropy is inexorably tending to increase; so in effect we are heading to our doom 'as time passes'. This implies that time is running all through the universe as an independent entity; but under relativity there is no longer a universal time. Nevertheless, according to the former editor of NATURE, Sir John Maddox, the very pillar of the scientific establishment," the reality of the process is beyond dispute". However the theory overlooks the logical fact that time is defined as 'a period of waiting', and that it is caused. Anything that causes a period of waiting is time. I use the tern 'processing' for our experience of time. Several agents are suspected to be causing what we know as 'time'---chemistry, inertia, delayed-reaction (such as is common in earthquakes, tsunami and other natural events), or even tugging from dark or invisible matter. Otherwise time does not exist as a physical entity, and Professor Eddington, too, has strongly castigated all those who still believe that time flows, as quoted in this book. So time as an independent entity passing by is logically untenable. Therefore entropy cannot be increasing 'as time goes on'. Time does not move; it is discrete and the procession of its units (like the year) add up to create the illusion or sense of continuity. It is even a surprise that the very scientists who swear by the Minkowski 4-D continuum insist that time is passing by as an independent entity, and that as it does so it causes increase in entropy (see the chapter 'Inventing Time' in Sir John Maddox's book, What Remains to Be Discovered.) He asserts that, "An ordered state spontaneously becomes disordered or the entropy is spontaneously increased...Does not behaviour of that kind provide an objective sense of the direction in which time increases that is independent of sensation in our heads?" The simple answer is no. Time does not run through nature; it also does not pass, only the cycles we use for tracking it are passing or moving. Time under relativity has no direction. There is no such thing as the direction of time; there is only the replication of time units---the one year is one unit of time and it becomes years as the earth continues to orbit the sun. The year does not need direction to continue. It is only one, and replicates to become many, centuries. Time is discrete because it all depends on the discrete yearly cycle; all the other units of time are fractions of the year, or are derived as fractions or sub-divisions of the year with points in association with astronomical features of the earth. The earth-year is the beginning and end of earth time.

It is common knowledge that scientists disparage philosophy. No one can blame them when even Sir Karl Popper has confessed that he's not proud to be called a philosopher due to the silly and irrational ideas now paraded as philosophical.[105]

Yet still philosophy is unavoidable, and the whole of science needs it. In his review of the book Science, Perception and Reality, Professor Herbert Dingle said exactly the same thing in defence of philosophy.[106]

Scientists are narrow specialists; they need the philosophers who survey the whole scientific field, from biophysics to astrophysics, the tiniest to the gigantic, the invisible to the massive stars so huge that millions of our tiny sun will find room in them; and they need these generalists to ask pertinent questions that the specialists may have missed, for we are in the dark about the universe and why we are here with the ability to ask such probing questions such as this: "If cosmic time is abandoned, then what is measured by the clock?"[107] What gives the human brain its incisive powers is the biggest conundrum of all. At the end of his article mentioned above, Professor Dingle had this to say: "...Science itself is rapidly transforming at least the external patterns of our lives, but those who practice it cannot reflect adequately on what they are doing simply because they have no time. That is an independent inquiry, and a most urgent one..."

---

[105] Popper did not mention names, but I know he's annoyed by the popularity of Wittgenstein because of his 'logical mysticism' and condemnation of physics which Russell said was his reason for rejecting 'the mad thinker' in the end. (Quoted from his book Modern British Philosophy, London, Secker & Warburg, 1971.). The phrase 'mad thinker' was coined by me because Wittgenstein was preaching a philosophy that puts an end to physics, according to Bertrand Russell. Any thinker who does that must be mad; and it says a lot about Oxbridge that they regard him as the greatest philosopher of the 20th Century. Outside science, technology and medicine, Oxbridge is not worth a farthing.

[106] Professor Herbert Dingle TLS 25th October, 1963.

[107] The whole of time, history and even existence, are implied in this simple question from the great English philosopher, Bertrand Russell. Nobody has yet even attempted a tentative answer, yet it's still relevant, and until it is satisfactorily answered, all discussions of time remain unscientific.

Yet, in my opinion, the present set up precludes that. Specialists advance theories consistent with formulas passed on by other specialists; if there are errors in them they get transported to other specialists to pass on till disaster or a major breakdown in communications brings a whole discipline crashing down. Often nobody gets hurt (the general public may not even hear of the problems in certain fields), because in theoretical matters, the distance between theory and practical effects can be so remote as to appear to be none-existent. Moreover there will always be proponents and opponents to any suggestion (the protagonists), firing their heavy guns at each other. But scientific research covering so many fields require the philosophers to knit them into a coherent story to satisfy the human soul and the specialist scientists cannot do it.

As a result it is extremely difficult to challenge any theory once it's been published, no matter how wrong it might eventually turn out to be. Time dilation is a notable example. It's supposed to mean time runs slowly with speed, but obviously that is not the case---at any speed who can read the speeding clock? Those travelling with it are also speeding with it and notice no problem. Before the dust settled on this conundrum, up popped the Minkowski theory of four-dimensional continuum, an idea so seductive that mathematicians just fell in love with it and overlooked its arbitrary nature. Succeeding generations of scientists have used the Minkowski formula to vitiate their suppositions and speculations about black holes. The fact that Einstein mentioned it in his general relativity is no excuse; we now know that he did not even bother to understand it. The answer, I suggest, may be found in the critical examination of formulas before incorporating them in new ideas, although I know that this is easier said than done.

But it's sad that whenever they try their hands at philosophy (some) scientists, though at the top of their fields nevertheless manage to utter such banal ideas as to make one feel sorry for them. For instance entropy cannot be marching with time leading to the death of heat, energy and the end of all activity as predicted in the second law of thermodynamics. There is no such thing as 'the march of time'. Our time is discrete---the year for instance---and so it cannot march. I don't know about the death of heat and all activity, but does that mean the end of the universe of all action and motion through entropy? What about the theory of frames? Are we to understand that our tiny planet is so pivotal in the universe that if the death of heat occurs here it will reverberate throughout the cosmos? Anything causing anything in this vast and unfathomable universe is just one instance of the multitudes of events and activities in the universe (positive and negative, forwards and backwards), and it is so vast and complex that only a fool would dare to make absolute statements about it one way or the other. As either individual persons or even en mass and together with our planet we're not even as big as one thousandth of a small ant's leg in comparison to the size of a cosmos consisting of about one billion trillions of huge stars as astronomers are now telling us. Many ideas about time are religious. The end of everything idea, when it occurs in any theory, has Heaven and the fear of death hidden in it or written all over it.

To those who are using these dubious ideas to claim that Einstein's theory of time is untrue, what I have done in this little book is to point out that there is natural time but it is caused, materially caused, and is not leading to the death of anything because it is not everlasting. Certain conditions spawn (or enforce) periods of waiting of various lengths on our senses; they also give periods of revolving lengths also implying time; and so for creatures with cosmically brief lives as we are, we can accommodate our lives within some of these periods of waiting—which we know as 'time', or time-span. Thus we use the orbits of the sun or atomic oscillations. Atomic time is also, like the year, our means of showing how time is passing and never what it is (I always have to emphasise this.). But all this is destined to disappear altogether when the sun decays. They have no more influence in the cosmos than a solitary drop of water has on the size and depth of an ocean.

Mistakes in science are almost entirely caused by formulaic thinking---or standing on the shoulders of giants. We've got to make sure they are permanent giants, useful giants, reliable and intelligent giants, not mere religious dreamers in exalted positions in the academe and elsewhere.

GRAVITY AND TIME 9

Scientists swear by the Bible that gravity causes time to slow down. Numerous theories have been built on this notion. But, again, it all depends on how the time is understood or defined. One basic thing in the world is that our time is based on the regular orbits of the sun by the earth.[108] This makes our time discrete---consisting of one year at a time, and the year too is pared down to the seconds and other units of time as fractions of the year, and therefore also discrete. That is the reason the units of time we have are all in individual units not in a chain---second, second, second and so forth. But, obviously, discrete time cannot be affected by anything---wind, rain, sunshine---except the earth itself. Technically discrete time does not exist as a living, pulsating entity. It's one moment and gone. What is confusing people is that the months and weeks and years are not temporary 'moments'; they last for various lengths of time. This is what we mean when we say planning time is spread over longer and longer periods. But the basic cause of time is momentary---the year is just one moment

---

[108] Other aspects of time have already been explained and others are discussed below---I am thinking of anything that makes us wait or involves us in a period of waiting like, ebb and flow, motion, growth, ageing, chemistry, silent time in sleep, etc.

although it gives twelve months for planning life's activities. It is momentary in the sense that it is determinate and has to be repeated to continue. As Professor Whitehead put it: "A time system is a sequence of non-interacting moments". It is a logical moment, although there may be several months in one of its moments---several minutes in an hour, several hours in a day and so forth. So long as the year is determinate and has to be repeated to continue, the time system derived from the year cannot be anything other than discrete time, with all the momentous implication of such a time system—it cannot march, curve, or cause the story of history, and so forth. It is also automatically secular and logically structured. Altogether, we are trying to conceive a time system to replace cosmic time and discrete time seems to fit the bill. But all theorists are fallible and cannot claim omniscience, except that what is logically unassailable cannot be rejected with illogical ideas. This is a perennial problem. We saw it in the discovery of special relativity. Many theorists could have discovered it or come close to it, but they're hampered by so many formulas relating to the eather, and nobody could tell anybody that the eather debate was logically flawed. Everybody agreed that electromagnetic radiation had to have a medium of transmission. Einstein had to ignore the eather completely to discover special relativity.

The theory regarding gravity and time, as stated by Professor Bernstein in his book, Albert Einstein and The Frontiers of Physics (Opp. Cit. page 110) is this: "In the absence of gravity, space and time are distinct entities. In the metric of special relativity they play distinctive roles..." This is what we want to hear because we live in a special relativity metric. The contentious issue is what he went on to state: "But in the presence of gravity the metric is altered, and space and time become mixed up with one another. The metric has four coordinates, but the space and time coordinates become entangled..." Yet, space and time are not different streams of many rivers to become mixed up, unless time is regarded as running through nature like a stream. The whole matter is hopelessly confused and confusing. If time is not synonymous with motion this cannot happen, and we know that our time is not synonymous with motion. It's not the same as 'Being' either. That would be a universal time but there is no longer a universal time. We construct our time. As I have argued above, we have had to do something to being (existence), or just being there, to get our time.[109] We could not have got this time without sentience or intelligence. For we know that time in the clock evolved from very crude attempts to the sophisticated Rolex watch. The result of this evolution is that we have come to realise that our time, being based on the repetitive yearly cycle, is essentially discrete---year after year after year, or second, second, second and so forth. Every second is a fraction of the year, so also is the atomic pulses based on the second. This time is not running all through the cosmos in the form of a thread that could be twisted by any force, gravity, or whatever. The time consists of units of time in procession. Therefore if the Minkowski attempt to equate space to time is rejected as arbitrary, then there is no way that time can be influenced by gravity, however strong.

Nevertheless, we are told that experiments have shown that clocks sent to space run faster than those left near the centre of the earth where the gravity is strong. If this is true then we have to interpret it as Einstein said in the case of clock paradox: one clock is working erratically because it is affected by acceleration.

In the end, we have to consider seriously how to do scientific research standing on the shoulders of previous thinkers. Innocently, the famous editor of NATURE was led astray by formulas. He thinks the process of time causing increase in entropy is beyond

---

[109] Almost everybody treats time as if it's the same as motion, 'being' or existence. This comes from tradition. For centuries that's what we believed and it was staunchly supported by the religions. Unfortunately, it's not true and we just have to learn the new theory of time as not even time but just how it passes by means of our physical cycles..

dispute. But a philosopher would ask, so what? We do not cause it. If under relativity time is limited to a frame and based on (constructed with) repetitive cycles so that it is necessarily discrete and not running through nature like a thread, then it cannot race with entropy to increase it or otherwise. It has nothing to do with it. The formula that it does overlooks the fact that time is no longer a universal entity. Time does not move. The repetitive cycles we use to try and trace what time is may move repetitively, thus giving us only discrete units of time---the years, for instance. But that can have no influence on the tendency of entropy to increase in nature due to unknown causes. Even then we are not so sure. The universe is too big and complex for us to know much about how it has managed to persist through deaths and renewals, despite these entropy increases. We don't know what time is so how can we tell whether it moves or not? The physicals cycles are what we count as the passage of time; so the cycles do move, but we do not know whether time is also moving with them. We simply do not know what time is; all we know is how to count the yearly cycles as the rate of the passage of time and use it to live in safety. Real time is caused by one or many of the agents mentioned above---but what it is we simply do not have a clue. So how can we tell whether it moves or runs through nature? Secular time bases time on the yearly cycle; if true, then what we call time is discrete and cannot move.

APPENDIX I:

TIME AND QUANTIFIED TIME OR THE PASSAGE OF TIME

We are all fond of using the word 'time' loosely to refer to the passage of existence in any form whatsoever. That may be called 'the unscientific' notion of time. In logic, science and philosophy, however, time is what Professor Richard Feynman called 'how long we wait'. This translates into the concept of 'how much time', or quantified time, so as to be able to tell how long we wait in mathematical language for universal application.

In any serious discussion of time, it does not make sense to just mention time. It may take centuries to understand that 'Being' on its own is not time; it's true that you have to 'be' in existence to know how to apply points to create intervals of time between points, but being on its own is not time; being has got to do something to nature (that is know how to divide it into periodic intervals) to create time---that requires sentience. *I insist that such a time system cannot move; what appears deceptively as the running of time consists of merely the motions of the cycles used for time.*

As discussed above, motion is not time either.[110] It shows time going, but the time will have been created elsewhere beforehand. At best motion is silent time; but I think all that is chemistry, for chemical processing can impose (or require) a period of waiting at the visual level, which is the same thing as time. The truth is that all sorts of things may be called time but none can show how it began. We can only rely on how we experience it, and that is through the use of points as applied to space to create time intervals. Logically, this is the best we can do either in science or logic. Everything else is sheer religious humbug.

So motion is not and cannot create the time we can mechanise into the clock; the simply reason is that it is multitudinous. But we can count cyclical motions and call each cycle say, 'a year'. If this cycle is continuous we can have years and years in perpetuity; and a year of course is time; it allows us twelve months to do whatever we want to do; it is also the standard measure of age.

And that, precisely, is what we do to get the time to programme into the clock. However this can only show how much time is passing by means of our physical cycles and never the real thing. We count mere physical cycles and call them the rate of the passage of time---but what is the true nature of time? In my opinion time is the same thing we call life. The secret of life is the same as the secret of what we call time, with the proviso that all we can ever know of this time is how it is passing by; therefore the implication is that we can never discover the secrets of life either. Obviously we have to be in existence; and we are in existence (I think therefore I am!). So it means we are living (we're there or here) and choose to use a regular motion as the rate of the passage of time. The cycle does not call itself time. It's mankind that regards it as the rate of the passage of time, or the measure of

---

[110] These are the two things (motion and 'Being') people normally assumed or imply to be time in ordinary conversations.

duration to guide his actions. That's all there is of time. It is not as scary as was previously thought; it's merely a device to guide our actions. One can even tap the finger to the same effect. Let's say a thousand taps means an egg is cooked or done; that's not different from saying ten cycles (minutes) means an egg is cooked. The earth's orbit is so long that we've had to sub-divide it into smaller units of time. But every second is part of the yearly cycle, and therefore logically part of a cycle. A second to go is not yet a complete year.[111] But deciding the when by means of the logical study of time brings in notions of earth time.

The nearest we can get to the definition of life is the logical definition of time, for the two are closely associated and don't seem to be separable. After many years of thinking it all came to me one day encapsulated into one word "When?" When is anything?[112] We can't have any existence without when (the time) it is or was in existence. Thus life is time and time is life, since we cannot define life without the "when?" it was or is there.

Ironically the logical definition of time reveals it as merely how it is passing by through the use of physical cycles. Thus a great conundrum, juggled round and round in the most exhaustive manner, becomes the greatest mystery. It is that time is life and life is time, simply because every second of existence is time; once you're alive you're expending time. It's not the same as saying it is the time allowed by God. It's slightly different; though I concede that the religions came close, very close. They've always had some back-room chaps bearing their intellectual burdens and some of

---

[111] Even nanoseconds are part of the earth-year. So are the atomic pulses used to mark time as they are based on the second---they merely lead to a more precise measurement of the second as a fraction of the earth-year.

[112] It's because existence implies time that the religions arrogated the right to decide what the nature of time was. Once time was liberated from religion, its logical definition was not difficult. The only problem is that existence still remains the attribute of time because we can only define existence in terms of the "when?" it's there.

them, like Kepler, were very good indeed. The difference between my theory and that of the religions is that I am asserting that time is life and life is time; on the other hand the religions claim that life spends the time already allotted by God. Also, while they have no proof of their views, I have logic and Einstein on my side---with Bertrand Russell as a providential bonus! Time is life and life is time because you cannot define anybody's existence without when he or she was in existence; the quandary is that this time is one of our own creation, or construction. So, again, we have to theorise on the basis that man has a hand in the logical definition of existence. Thus the fields where man the observer cannot deal direct with reality are now three: Plato's simile-of-the-cave, the Einstein 3+1 formula, and our present definition of what we mean by existence---that nobody can exist without his when (time) of being there, but we construct this when ourselves out of the parameters we find in our environment. By the way, the Einstein 3+1 formula is included because of the time element---and we create the time---so it means we contribute to the nature of physical reality as perceived. That is to say, we determine physical reality from the three aspects of space and matter plus time, our time. This is the reason Minkowski incorporated the time element in his equation so that we can write $S=CT$ to represent all physical reality and it is for the same reason of mathematical economy that mathematicians insist he is right. Unfortunately intention (or human desire) and physical reality are poles apart.

Again, my theory means your very existence is convertible to time the moment you are born not that the time is what is permitting your life to endure. The when of your existence comes in as soon as you're born and never stops---time is unavoidably continuous. Once a person is alive time takes over the control of his or her life in the following manner: you're alive. To continue to live on you've got to live strictly in accordance with the earth's motions and environmental conditions. These are what have been converted to time, based on the earth's motions. This time is unavoidably continuous because the earth never stands still. Thus from birth a person is controlled by the motions of the earth as we have converted them to time units---therefore time controls life. To be is to be part of the yearly cycle or be spending part of it---you cannot be without spending time. Nobody can exist without the 'when' or time of his or her existence. Life and time are inseparable. So to establish that time is secular is, for me, the greatest philosophical intuition or insight. I think it solves the last conundrum about time and life too.[113]

Deciding 'how much time' by means of regular cycles is the main job of the interpreters of time. The context of any proposition (in science, mathematics and philosophy) must always show or imply the sense of 'how much time' in it, or expressly show the quantity of time proposed. Of course, time may pass when one is not conscious of it. But in all cases, when one wants to know how much time has passed, or will pass (as in futuristic propositions), mathematics must be used to quantify the time. And let me stress

---

[113] The conditions conducive to life have been reduced to time units; these are based on the earth's motions and the earth never stands still, so every second is part of the earth's motions, and these seconds are oppressively continuous. So the when of a person's life is a time unit, without which he cannot live because it is tied with the earth's habitable conditions, without which the earth will not be habitable. Thus if you do not live according to the seconds (or the motions of the earth) you'll perish. Life is therefore linked to time, or life is time and time is life. The whole theory is based on the fact that the earth's motions are continuous. It means you are free to live for one second because the earth's conditions have approved that it is save to live for just that second. With forward planning things are easier than that, but the principle is the same. You live because there is time for it allotted materially by the earth. If you have a second, a minute and hour or week to live (without medical conditions, it means if you are to live at all at any moment or time) it is because the earth has indicated that it is safe to do so, or that it'd last long enough for that; if it were not safe to do so the time won't be there, those time units won't occur. So we live by time allotted, but not in the religious sense. However, the religious chaps were very clever, as we arrive at the same conclusion by logical reasoning.

again that we quantify time by the use of external cycles in union with any sense of duration of anything whatsoever.

Quantified time is 'time in a clock', any clock at all. And the clock, any clock, can only show time as independent of space. Space-time is automatically quantified as it is derived from space with points, which is the only reason for calling it 'space-time'. Discrete time can only pass through the succession of the individual units. On this point, Leibniz was absolutely right when he said time is succession. What was lacking in his day was the concept of discrete time; with this new concept in our post-relativity world, we can now see clearly as to how time passes and seem continuous through the succession of its separate and individual units: second, second, second. Plus the hours, weeks and months all the way to the year, which also passes in the form of year after year after year.

It may seem surprising, the springs of a thousand legends, giving rise to supernatural speculations, that we have an extremely ingenuously smooth time system, so cleverly structured that it is there when we are born and there as we die, and always passing by. For this reason we know that "Time does not wait for anybody". Scrutinised under a logical gaze, however, time is not so rosy; it is only one moment, repeated to pass by and seem continuous so that arithmetic can be applied to its accumulations.[114] This, as we know well, happens when we reckon time for futuristic planning, and backwards as in historical narratives.

---

[114] Let me explain that space-time is necessarily discrete. We have only recently come to understand space-time from Albert Einstein; yet time has always been discrete, consisting of only one unit (or moment) of time---of whatever length. For there is only one year, and all other units are obtained from the year in the form of separate units of time. To get more years we repeat the one year exactly. Thus we have second, second, second; or minute, minute and hours and so forth. Each is a moment (or a unit) of time in its own right.

But for the union between the sense of duration and external cycles giving us units of time out of the moments of time, time for the clock would not exist at all. Presently philosophers see time as rather a straightforward pragmatic entity, albeit not as simple as it is normally supposed. It is partly a confidence trick that makes the clock work continuously, the trick of continuity is in the repetitions of the seconds, or of the units of time, all of which are to be understood as single moments---the realities---of quantified time. It is also partly physical (using physical cycles for the process of quantification); and partly philosophical, i.e. according to Einstein without time physical reality is indecipherable, or cannot be properly (accurately) determined, hence his equation for motion consists of the three spatial coordinates plus time in the 3+1 formula of physical reality. This, of course, is contrary to the Minkowski formula and I think this is much more scientific. It is true there is an element of subjectivity in it because the time is man-made, a human concept 'constructed' by man. But at least it is not as arbitrary as the Minkowski imaginary time coordinate.

To sum up, we have to recall that Einstein made man the observer part of the observed. Plato also made man part of the observed with his simile-of-the-cave notion of perception, meaning we perceive the external world not as it really is but just how we are made to see it. When it comes to time (as the most important aspect of life) the situation is the same. We see the world and time not as they are but just how we are put together by the human architect to see it. Logic, of course, is our principal instrument of perception, theory and knowledge. Thus Bertrand Russell, as the great logician he was, summed the Einstein theory of time up and concluded that cosmic time should be abandoned since it cannot logically account for the nature of time as discovered in experiments. Professor Eddington also concluded that those expressing doubts about the Einstein theory of time were making meaningless noises. The founder of astrophysics was convinced by the secular theory of time. One reason, as I have pointed out, is that the religious notion of time and the nature of time discovered in logic are pretty similar: we don't know what it is but all of us accept and live by the yearly cycle as the passage of time, and have been doing so for centuries---centuries which are just the number of times the earth had circled the sun. Any good logician would sense that the true nature of time was not far away, especially after Russell asked the most important question about time---if cosmic time is abandoned, then what is measured by the clock?

Unfortunately, researchers did not follow this logical trend to try and discover the true nature of time, but rather jumped on the Minkowski bandwagon to propose concepts of time travel as 'a scientific possibility'. It is a sad reflection on the mentality of some writers that they should seek to twist the mind of mankind to concentrate on The Afterlife rather than the actual physical reality influencing human life now.

I've always felt that if this had not happened (with numerous books about time travel selling millions while contrary suggestions are rejected), man could have done really good researches about the nature of time. Minkowski and Kurt Godel bear the blame. But that is not all. Man is basically more interested in life after death than anything else. Well, if the Minkowski formula for equating space to time is not logically valid, it means it cannot happen, and if so then travelling by space-time is not feasible. We have to go back and research time as a secular entity that is separate from space exactly as Einstein made it in his special theory of relativity. At that stage the Russellian question comes up again---what is measured by the clock? Let us consider this question in the next chapter.

But I must stress that discussing the passage of time is necessary only to accord with popular ideas of time; otherwise I don't think time is ever in motion. To me the reality is that the cycles we use to reckon time make us think that time is running all through the cosmos. Yet time consists of separate moments, no matter how long each moments happens to be; so it can only advance through the replication of the units. Our time is based on the earth year which is so long that we've had to sub-divide it down to the seconds---but the main unit and its fractions advance by replication, not by running. That idea is a mental deception.

WHAT IS MEASURED BY THE CLOCK?

Our time is based on the repetitive orbits of the sun by the earth, and evidently the earth never stands still.[115] If ever it does stop going round the sun, our time system will be completely nullified; but, of course, life will go on. It is inconceivable that all life will be extinguished instantly the moment our time is (mathematically) nullified in that only quantified time would be lost. This is the best proof there is that life is not based on "time allowed", as the religions believe; rather time is a union between the sense of duration and external cycles---therefore man had something to do with the time we have in the clock, the only reliable time, as quantified time.

All the religions speak of "time allowed" for the duration of a man's life. They had to, because the nature of time is easier to explain as a providential bounty than anything else. To be honest, without a cosmic explanation for time, what is time; to put the question in another form, what is the origin and essential nature of time? Everybody believed that it's divine until Einstein and Lorentz found that it can begin from anywhere. Of course, it is assumed that the clock measures time. Even Bertrand Russell talked about measuring time, asking what is measured by the clock-

---

[115] The orbit of the sun is what we find most convenient to use as a measure of the passage of time. It's no accident. All the features of time are features of astronomy---events that happen to our planet and affect our lives. Hence the day and night system taught us how to keep time, the moon's phases and the seasons resulting from the earth's positions round the sun did the rest.

--but from where? And what is it that the clock measures?[116] The clock maker will say he invented the clock to reckon time in the sense that everybody knows---but what is that sense of time? The mechanics (or clock-makers) used the day and night system, the moon's phases and other astronomical features of the world as far as they're concerned, just to help us measure our version of a universal time. In other word, they're just as ignorant of the true nature of time as everybody else.

When it is postulated that general time permeating the whole cosmos (and therefore the same everywhere) does not exist, the first implication is that every 'body' (or planet) has to have its own time; it is not coming from the cosmos therefore it must have originated on this planet. So let's find out how it all began. That is the first implication. The second is that, as a result, cosmic time is abolished---although it sounds tautological, it still has to be emphasised, as well, and most clearly because the 'cosmic time instinct' is permanently ingrained in the human mind. One reason is that time cannot be suspended; but the more cogent reason is sheer intellectual incompetence plus fear of the unknown. We are always using it, and so it does not make sense to just say that it is not there. But if it is there, and did not come from the cosmos, how did it begin? And the obvious fact is that it is *always* there. Even before we are born, and also as we die to leave it behind. Yet it cannot be supposed that each body's time is a version of something

---

[116] Without the explanation that what the clock measures are cycles of duration, or duration reduced to cycles, metaphysically interpreted as a union between duration and its conversion to external cycles, time can never be logically accounted for. We will just go on using it---but in what form? In the form of units (year after year after year, and all the seconds and so forth derived from the year); yet that means the same thing, namely, a union between duration and its conversion to external cycles. For the year is only a physical orbit of the sun. It is not time. It is the practice of humankind to call it 'a year'. We use it as our basic unit of time, as a matter of convenience. Otherwise in nature it is not time. As a matter of fact, we can use something else---we can tap the finger, for instance.

'naturally existing', whether it permeates the whole cosmos or not, with the necessary but illogical (little 'academic') proviso that it may not be the same everywhere but varies with individual bodies in accordance with unknown natural laws.

It is plainly evident that this erroneous sense of time dominates scientific thought. Hence time is not defined in physics; and as a result, the Minkowski fiction makes sense to some scientists. They just say "as time goes by". Only Professor Arthur Eddington has redeemed physics by warning that it must never be forgotten that the 4-D geometry formula is "fictitious and arbitrary"---but they have chosen to ignore him, partly because Eddington was afraid to mention Minkowski by name, or maybe he's just cunning. The era was incredibly sensitive: There was Einstein, Planck, Russell, Whitehead, and the mass of aggressive no-nonsense mathematicians who regarded Minkowski as the genius who made relativity accessible to scientists. Thus Eddington had good reason to be cautious---nevertheless, since then everybody refers to the concept of space-time as 'artificial'. Let me explain another small point about original ideas. Even the originators do not stick their heads on them, because they're never absolutely certain that contrary ideas would not emerge; and we all know that most of the time they've emerged to shame cocky theorists. So even Eddington might have been a wee bit afraid of the pure mathematicians---even Newton was, and if David Hilbert is to be believed, then Einstein too was!

Thus Russell's query is important, namely, "If cosmic time is abandoned, what is really measured by a clock...?"[117] My answer, of course, is that outside the union between the sense of duration and its conversion to external cycles, time does not exist to be measured.[118] The very act of 'measuring' is the time in essence---

---

[117] ABC of Relativity, Ch. 4.

[118] I believe the origin of the sense of duration is in the brain; that it arose from how the

like moving from point to another point, time is going, so that time becomes 'relation between points', or intervals between points. The cycles are time units (the years, for instance), and the time units constitute the time: a year is a cycle, but it is our time, the basic unit out of which all other units are derived.

However, the cycles are the creation of man for the sole purpose of converting the sense of duration (of anything or any event, like the period it will take to reach the village from the farm before nightfall to avoid predators), to his time units to guide his activities. So the clock does not measure time; it rather reproduces units of time programmed into it by the clock-makers. It should be remembered that the seconds are put there by the clockmaker; but where do they come from? The answer is that they come from the subdivisions of the year. Otherwise the time does not exist anywhere to be measured---the units constitute the time. Without the year there will be no seconds, and the like, all of which are derived as subdivisions of the year. As hinted above, you can even dispense with the year and its subdivisions and tap your finger, if you will not get tired. A million taps means it is time to go to bed, and so forth; outside the units of time, time does not exist to be measured; but the units are the creations of man as quantified time to record the passage of existence in his experience in manageable units for cultural purposes.

I conclude that what we call time is the mere physical manifestations of it that we use (as periods of waiting) to organise our lives. These manifestations (or physical cycles) that we call 'time' are in motion of course: we count the earth's orbits---caused by motion---as 'years'. That process has been taken as the march of

---

brain was put together---generating the sense of time-lapse---and grew with it. For the brain was not formed at once; there must have been lapses of time and that created an instinct buried deep in the brain and it affects us. This is only speculation but gives us something to think about.

time. But we don't know what time is to tell whether it is marching or not. My feeling is that events do march and they have time associated with them thus misleading us into thinking that it is the march of time. In fact we can never know what it is, if it exists at all. The causes of the cycles we call time units may be chemical, inertia, dark matter, momentum, motion, kinetics, delayed-reactions and so forth. They cause what we experience as time. If real time does exist we cannot know it because it is shielded by the parameters we use for time. This echoes the Platonic Simile-of-the-cave again---it seems man just cannot perceive real reality. To me that's not so strange, for we are so insignificant anyway. The real surprise is the incisive power of our brains.

APPENDIX---II

THE PRINCIPLE OF MATHEMATICAL EQUIVALENCE

In nature there is reality and our perception of it. I subscribe to the Platonic simile-of-the-cave theory of perception. In the word 'perception' everything man does in life is implied, including mathematics, since we can only act by perceiving the true nature of the physical world; I am using the word in a sense akin to 'experience'. The problem is pure mathematicians normally are permitted to imagine things to satisfy their nostrums, so that they do not rely solely on their percepts alone. However outrageous, they can defy reality, even gravity, logic and common sense, and leave it to the applied mathematicians, to find out whether what they have assumed is really there in nature, so that their theories based on them can be seen as true or not. In no other profession is this sort of thing allowed. Even one of the greatest mathematicians Britain has ever produced, Professor Sir Arthur Eddington, criticised that common mathematical tendency in his book, *The Mathematical Theory of Relativity*. I have quoted him above in the text, but it will do no harm to repeat it as it is vitally relevant here. He said: "The pure mathematician deals with ideal quantities defined as having the properties which he deliberately assigns to them. But in an experimental science we have to discover properties not to assign them..." The principle of mathematical equivalence should make them think of the practical consequences of their imaginary properties, although I doubt it, but that is another matter. The rule is that mathematicians should not seek to make the basic features of nature what they are not quantitatively, or cannot be physically; any such propositions are bound to falter. Note that we are talking only of basic phenomena. By the very nature of man, it seems everybody can make qualitative/physical changes in peripheral nature not quantitative changes in the fundamental aspects of nature, and time is the second most fundamental feature of both nature and life.

The principle means that, in effect, one cannot use mathematics to state, say, that there are ten trees in a field, and propound theories about them if, in actual fact, there are only two. This is slightly different from assigning imaginary properties to nature. It is different because it relates to 'quantities'. Six into four won't go, or something like that. The principle of mathematical equivalence rules that, to accord with physical reality, one can only talk about two trees, or as things are not as the mathematicians want them to be. Nature is not there for the convenience of mathematicians; it is neutral. That was the advantage we gained when the ancient teleological interpretations of phenomena was discredited. Therefore this rule is not to be scoffed at. I regard it as one of the strictest doctrines in logic, metaphysics and science. Science means logical thought in physical applications; metaphysics is logical thought in abstraction and mathematics is logical thought by means of symbols rather than language to facilitate the handling of size, weight, distance, volumes and complexities. So all four disciplines (including logic) are inter-relater.

It is not often realised how progressive is the study of philosophy. Quietly but surely, many entrenched myths from our primitive past are being discredited one by one by philosophers. One of them is teleological argument. With that and many other ludicrous intellectual fashions out of the way, it is unacceptable to regard any concept as 'compounded for the convenience of the mathematician', as Russell defined the Minkowski theory of space-time. Someday, we may get scholars writing about the many myths philosophers have discredited through their quiet researches to foster science and progress generally. So I regard this principle of mathematical equivalence as a strict and necessary doctrine to prevent mathematicians arrogating the power and right to alter nature quantitatively in the fundamentals of physical reality. We shall, and should, continue to alter nature qualitatively to our benefit---gardens, buildings, roads, cities, waterways, canals, railways, bridges, tunnels, all science (bar destructive devices), and all art, sports and so forth. They do not change nature but beautify it; but quantitatively, never. We cannot make one object two, or two objects one, physically. It is not possible realistically. Not in reality only in the imagination; to jump from the imagination to live conditions can be dangerous.

The origin of the rule will help the reader to understand it well when spelt out: it occurred to me when I was pondering Hermann Minkowski's claim to have made time and space into one entity as from the moment he outlined his theory, as previously quoted, in the following outrageous (even cheeky) statement: "The views of space and time which I wish to lay before you have sprung from the soil of experimental physics, and therein lies their strength. They are radical. Henceforth [that is, from the moment of his lecture] space by itself, and time by itself, are doomed to fade away into mere shadows, and only a kind of union of the two will preserve an independent reality". This is to combine two things in nature into one with mathematics ('a kind of union of the two...') It means he knew they were two independent aspects of nature. How could he have made them one from the very moment of his lecture? (Yet mathematicians continue to accept his formula as true.)

He spoke of experimental physics. In fact, the only experimental evidence pointed to time being 'local' in nature; and Einstein adopted it in his special relativity; it's the Lorentz t^1. There was no suggestion that time had been found to be inextricably intertwined with space---rather the suggestion was that time could not be had without space; and that once you have space, you can create your own local time. What Einstein did was to interpret local time to mean "The only Time" we can have.

The actual physical reality known to be in existence was precisely as Minkowski himself stated it---namely, that time and space were two separate things. But it is interesting that he sought refuge in experimental physics. In that sense he did not breach the principle of mathematical equivalence. It shows that he was really a very good thinker; he had to be that good to convince Einstein to adopt his formula for general relativity, which came ten years later. The unfortunate thing for Minkowski and his followers is that the evidence he cited was really irrelevant to the claim he was making. He needed physical support that time and space are inextricably intertwined and therefore constitute one entity. The evidence that had been discovered by Lorentz and Einstein was that time was essentially local in nature, leading to the supposition that 'there are as many times as there are bodies', and that, additionally, time is different in different places, and also under different conditions. The principle of mathematical equivalence can be used to refute Minkowski's claim to have made them into one entity as from the moment of his lecture.

The rule stipulates that he could only have spoken about time and space as they actually were in physical reality, which, he admitted, were two separate entities. The reality before Minkowski was that there was space, and there was time. Even the great Einstein himself made them independent in his special theory of relativity. So it did not surprise me that Professor Sir Arthur Eddington and Bertrand Russell described the Minkowski proposal as arbitrary and fictitious. However, it did surprise me that mathematicians ignored this strong condemnation to claim that they could not understand Einstein's ideas without the Minkowski fiction.

That made me sit up and think, think of a principle to require mathematicians to relate their suppositions to exactly the nature of physical reality laid out before them, not as they would wish it to be to accord with their nostrums. I came to the conclusion that mathematics can only mirror reality, *not to alter it with mathematics alone.* So the principle of mathematical equivalence is this: Mathematical statements (or equations) must strictly accord with physical reality. That is the true meaning of the term "equation". It means no mathematical quantity can exceed or reduce what the actual physical quantity is. No mathematics can make one thing two, or two things one, without physical divisions and unions. Minkowski failed because, as Professor A.N. Whitehead has pointed out, time and space *still* pass through nature as two entities, not one.

APPENDIX---III

WHY SPACE ON ITS OWN IS NOT "SPACE-TIME"

In Einstein's special theory of relativity, we learn that, "In the absence of gravity, space and time are distinct entities. In the metric of special relativity they play distinctive roles."[119] Nothing in special relativity has changed since then to make all space "space-time". Yet in all their suppositions cosmologists and astronomers always refer to space as space-time.

Let me set out the facts as they are at present, as argued all through this book, and hope they will see the light. To begin from the very beginning, the whole idea of space-time comes from H.A. Lorentz; until then space was space and time was time. It is true that in special relativity Einstein made space and time dynamic rather than the Newtonian absolute; but being dynamic merely means they are changeable under different conditions. But about time alone Einstein avers that he was able to complete special

---

[119] Professor Jeremy Bernstein, in *ALBERT EINSTEIN: and The Frontiers of Physics*, Op. Cit. p110

theory of relativity five weeks after he gained the insight that the Lorentz idea of 'local time' can be defined as 'time, pure and simple'. So let us examine the Lorentz notion of local time.

H.A. Lorentz found that time runs slower when in motion, known as "the dilation of time as a measure of moving clocks". He could not understand why and literally put it aside. He called it 'local time' or t^1. To him it was not 'the true time' but a mathematical auxiliary or curiosity---not very important. Time, he said, was time, denoted with t, and t^1 was something you get as your local time, but certainly not applicable in the outside world as time, because it was a mere mathematical curiosity. May I remind the reader that all this has been given in detail in the text above? I have even mentioned Lorentz's own statement that he thought he failed to discover special relativity because he did not regard time dilation as important.

Strangely, however, as one of his brain waves, Einstein worked this into his theory of frames. The dilated time was 'local time'---the time of your locality. Now, if the universe was fragmented, then local time would be somebody's time, which to him would be running normally like any other time, but to outsiders would be running erratically (or slowly, in this case.)

In actual fact, that was the case with the Lorentz discovery. People outside the moving clock saw it as running slowly; but those carrying it in the moving vehicle noticed no difference in its performance. That is the genesis of the Einstein theory of frames. Otherwise time was separate from space. What you will find is that it varies under different conditions, simply because everybody has to have his own 'local time' in his locality or inertial frame. But since time is continuous, and having made it a separate co-ordinate in the study of phenomena, dynamic space would have different time co-ordinates at every turn. We recall that Bertrand Russell has stated that from the sun's point of view the tram never repeats a former journey---because the time co-ordinates would be different. Since time is a separate co-ordinate in the determination of physical reality, different time co-ordinate implies a different situation, different physical reality.

This was the state of affairs when Hermann Minkowski came in with his theory of 4-D geometry making time part and parcel of space---all space. So that cosmologists and astronomers call his theory "The Minkowski Universe", meaning that all nature is subject to the 4-D geometry, where time and space constitute one entity. But let us swiftly add that the foremost mathematical interpreter of relativity was our own Professor Sir Arthur Eddington, the man who confirmed the general theory of relativity. He wrote the definitive book on relativity, called *The Mathematical Theory of Relativity*. About the Minkowski 4-D Geometry, he stated clearly on Page 9 (Ch. 1.1.), as already quoted, "*Such a mesh-system is of great utility and convenience in describing phenomena, and we shall continue to employ it; but we must endeavour not to lose sight of its fictitious and arbitrary nature.*"[120] He was not the only great mathematician who described the Minkowski formula as arbitrary. Bertrand Russell also said it was based on arbitrary assumption. He made it plain that because of that the derivation of the Minkowski 'interval' as time from space was not valid.

Let me try and explain again the reason mathematicians still adore the Minkowski theory---even though they know that it is fictitious. It makes things easy for them. Yet it is not true. They accept the novel Einstein notion that time must be made a distinct co-ordinate in the description of phenomena. The problem is that at

---

[120] The emphasis is mine. I have had to mention this several times, because, quite honestly, I am outraged by the mathematicians' desire to perpetuate the Minkowski formula as if it is really true of physical reality---yet it is not, and they know it. At least one of their own numbers told them so.

the same time Einstein made all time (any sort of time) 'local time'---the time you create for your own local purposes, as Lorentz had discovered. Einstein extended the Lorentz idea to all nature. With the universe being fragmented, it was impossible that one system of 'dynamic time' (as opposed to 'absolute time'), could apply with equal validity to all fragments of the universe. As a result he said there are as many times as there are bodies in the universe. Nobody can contradict Einstein on this matter. But mathematicians found that creating your own time to add to phenomena to acquire concepts of physical reality puts too much power in the hands of mankind. (I suspect there are religious sentiments in this.)[121] Besides, it was complicated. The Minkowski system was easier;[122] you just have to mention the Minkowski space or ds^2 and move on. It comes with time already embedded in space as part of it---so the whole of space is 'space-time' and every time is also 'space-time'. The caveat of Professor Eddington was quietly ignored. Soon everybody forgot about this; Eddington and Bertrand Russell were dead; and there was nobody clever enough to notice the discrepancy and question them about it. Of course, that leads to a distortion of relativity, but mathematicians are the arbiters of truth in mathematical physics and they were the ones benefiting from the Minkowski theory, and therefore preserved it.

---

[121] The Minkowski formula makes time universal again after Einstein namely, as something in general existence mysteriously (harking back to Pythagorean mysticism in mathematics), which can be invoked with the appropriate mathematical symbols; not as something you create in your own local space with the application of points to space, which makes time completely secular. It seems to me that humankind is not ready to accept time as purely secular. Those of us who have already made the necessary psychological adjustments for accepting time as plainly secular are not regarded as normal.

[122] It was difficult in mathematics but easy in logic and philosophy; and let me hurry to add that, because of the involvement of time, the whole notion of local time or space-time has philosophical implications, since time is the second most important thing in the world, second only to life itself.

Otherwise it is not true that all space is 'space-time', while all time is also 'space-time'.

Yet it is true that time is always space time. You cannot have time without space; not because the space comes with time inside already, but because all time is known and used in units and units only, which can only be had by the application of points to space to create the time intervals as "relation between points". There are elements of time in the mind as the internal sense of time, known as the sense of duration. But we have got to link duration to external cycles to give us usable time in units, as I have explained above. For example, without space we cannot have the year; yet the year is our basic unit of time out of which all other units are derived. This brings a little complication but nothing serious. The reason is that you can only create time, as 'intervals', or as 'time units', as I suppose (because the year is only one unit of time and we derive all other units from the sub-divisions of the year with points or mathematics), with the application of points to space, thus making time a product of space, and therefore 'space-time'. The truth of the matter is that you cannot have time without using points to divide space; it makes time necessarily discrete, being the product of points. Therefore time is always 'space-time, or properly '*space-timed*'. But that is all the connection between space and time, except that space is required, again, for displaying time in units as we have in the clock.[123] The clock, any clock, does not give

---

[123] The poignant question posed by Bertrand Russell comes up again, namely, in the absence of universal time, what really is measured by the clock? (ABC of Relativity, Ch.4.) This is a very serious matter, because if cosmic time is abandoned, there is no time, or any logical explanation for the time we have. The answer, of course, is that the clock does not measure time. It is deliberately programmed to *reproduce* specific units of time: second, second, second, leading to minutes and so forth, to accord with the cycles of the earth, so that about 31,536,000 (or so many) seconds will coincide exactly with the earth's orbit of the sun, called 'one year'. To have more years, we go round the sun again and again and again---hence perpetual time. Units of time in procession give us continuous time. From the Einstein concept of space-time we know that time, since it is produced with points, has got to be wholly discrete.

'flowing time'. It merely reproduces units of time programmed into it. The old mechanical clock based on coiled springs gave the best illustration. The springs are manufactured to release units of time: second, second, second. If one failed to rewind the springs, the clock stopped ticking. The springs provided the clock's energy, but were strictly programmed to reproduce time in specific units only.

After the time is derived in this way, it becomes separate from both the space and the points used in creating it. That is why Einstein made them separate entities in special relativity. For, apart from the condemnation of the Minkowski 4-D geometry which assumes that time and space constitute one entity by Russell and Eddington, Professor A. N. Whitehead has also pointed out that time and space *still* pass through nature separately---not as one entity. To add to these, I have humbly suggested the Principle of Mathematical Equivalence above, which can also be used to denounce the Minkowski arbitrary and fictitious formula.

APPENDIX IV

THE MISCONCEPTIONS OF TIME IN RELATIVITY

It must not be supposed that the problem of time in relativity has been conclusively settled. Relativity is physics. When a problem is solved in physics the solution is always clear, precise in mathematics, and universally applicable; but time in relativity at present is very vague, neither definite nor precise, not least because consideration of time is a philosophical inquiry, and a very serious one too.

The arguments here are that the original Einstein theory of time can be used to solve the passage and continuity of time. Unfortunately, Herman Minkowski made the question of time in relativity immensely complex and vague, not at all like the original notion proposed by Einstein. Indeed, as a result, the question of time on the whole is destine to keep the philosophers busy for several centuries as their nostrums become footnotes to Einstein instead of Plato. As regards the physicists and cosmologists, as opposed to the philosophers, they believe that the Minkowski theory makes things easy for them; the problem is that it is just not true of the physical world.

Bertrand Russell has said the concept of space-time is perhaps the most important theory Einstein introduced. To me, there is no doubt (no 'perhaps') about it. It is the most revolutionary theory in human history simply because time is second in importance only to life itself---and yet that life cannot even be lived as a well-organised existence without time. That is how momentous time is in human affairs; and Einstein has shown that it is very different from what it has been traditionally assumed to be. Secondly, he insisted that it should be taken as a separate coordinate in the study of phenomena. In the determination of physical reality, because of Einstein time is a co-ordinate in its own right just like the height or length of matter and space are, thus making Man, the observer, part of the observed, since he has to add the time in the 3+1 formula. Those mathematicians who assume, on the Minkowski theory, that time can be incorporated into space with mere mathematics so that we can dispense with the 3+1 formula and the metaphysical role of man in the determination of physical reality, are contradicting Einstein, which is something approaching a hanging offence in science. On the contrary, it is possible that the passage and continuity of time can be conclusively resolved with the original Einstein theory of time as space-time, or local time.

There is obviously fear in some quarters that time cannot be something we invent by ourselves. Of course, if 'there is no longer a universal time' we have to find out how we get our time.[124]

---

[124] It is not often realised that philosophy is of great importance to science; and, as an example, this is the sort of thing philosophers do behind the scenes to make their suppositions indispensable to science in general; for the philosophers service every branch of science. The phrase 'survival of the fittest' from biology which has passed into general usage in science and linguistics, was coined by a philosopher, not Darwin. All the sciences need philosophical interpretations. In the quotation above from Professor Dingle, he was saying this very strongly in respect of physics; but all the sciences need the same

However, nobody is claiming that man invented the whole of time. Rather we have found that we invented how to quantify time by linking the natural sense of time as duration in the mind to external cycles. This sense of duration of anything is obviously connected with the memory mechanism for the retention of images and concepts in the mind.

Let me stress again, and more strongly, that the sense of time is duration in the mind. In his *Mathematical Theory of Relativity*, Professor Eddington made this absolutely clear, as quoted above; and we have got to take that view seriously because the theory of time outlined in this book is based on relativity. Unfortunately the mental sense of duration is not enough. It cannot give time for general use because it is private. The word 'time' is meaningless until it is objectively quantified. We need time in units to apply to the external world---i.e. to mechanise in the clock for general use, so as to be able to tell 'How much time' at a glance---see Appendix I above. This is achieved with external cycles, the most basic of which is the earth-year out of which all other units of time are derived with mathematics. And it is maintained that this is in complete conformity with the Einstein notion of time, and therefore incontrovertible. Above all, it is the only means by which we can logically solve the problems of the passage and continuity of time.

For now, we are told in all earnestness from the discussions above that relativity is not properly understood. This may be so. But actually relativity is only a theoretical system, a suggestion. It is based on the suggestion that physical reality is not homogeneous but fragmented, and therefore subject to different natural laws. This applies to both special and general relativity. Bertrand Russell called it 'a logically deductive system'. In plain language, 'a new philosophy of physical reality' so logically structured that it demands attention, respect and serious study. And these Einstein
___
sort of assistance from philosophy, including mathematics and logic.

has certainly achieved. With Einstein alone we are not talking about genius but a godlike intellectual phenomenon never seen on this planet before; he reconstructed the world of physical reality single-handed, that is the reason he is indispensable to both scientists and philosophers.

So Bertrand Russell was absolutely right. Einstein's system is a new logic of physical reality, and it works. But theoretical physics is most unlike the physics we apply in laboratories. Ordinary physics is much more like chemistry; it has consequences. The Nobel Committee was right to award Lord Rutherford the Prize for Chemistry, even though he regarded himself as a physicist, who had rather cheekily claimed that "all of science is either physics or stamp collecting"!

In theoretical physics there are no obvious consequences, so it is difficult to judge the merits of suggestions. Instead, when we get a new theory in advanced physics (rightly or wrongly), three things will happen. I mean, all three will definitely happen in succession, whatever may be the merits of the new proposal. First, we will get interpretations of the basic theory proposed in such complex settings (or confused formulas) from rival theorists that the debate just has to go on; nothing will be settled in the meantime. But because there are no consequences, nobody will get hurt, no machinery will fail to function; avoidable calamities will not occur. The rains will not stop; the sun will not dim.

The most recent example was the eather debacle (or debate). Secondly, we will get accusations and counter accusations of misrepresentations and misunderstandings. The third possibility (because philosophers share with theoretical physic one subject-matter, being the determination of physical reality), will be philosophical interpretations to arrogate the almighty right to shame and discredit some of the factions in the debate, only for philosophers of different schools to turn the tables---and so the debate will be carried on and on. These philosophical discourses are often pretty profound, giving several intelligent interpretations without being able to settle the argument one way or the other. Strangely, that is how we eventually acquire our knowledge of the external world, sometimes referred to as the practice of 'academic freedom'. That is what happened to Plato. And that is what is happening to Einstein as he has come to replace Plato, in fact, to make his basic suggestion redundant, if not completely false, due to the quantum theory.

A careful examination of what has happened to Einstein's theory of time so far betrays elements of all three conditions. First, we are told that 'most definitely' due to Einstein's analysis of 'Order and Simultaneity' there simply is no 'standard or absolute time frame in the universe'. ('Time Frame' or 'Time Reference' means the same thing. It means the logical criterion of validity.) This is generally accepted as true; for it is reinforced by the Lorentz time dilation and local time concepts.

However, it implies that time in the abstract is utterly indefinable, as I have shown above with discussions about the earth-year. The year is indefinable; other time units in use on earth are defined in reference to the year. But the year on its own is logically indefinable. Again, all time units, down even to the cesium units, are based on the earth-year; they are meaningful only as related to the year; but like the years, on their own (that is in the abstract), none of them can be logically defined. How long, for instance, is a second in logic without reference to something else? The result is that we all have to use the clock, or clocks, based on the earth-year. By this theory of time (as quantified time), the human intellect is built upon the concept of "points and instants". Instants do not exist independently in nature. Only points do; they had to be discovered by man, but they do exist in nature independently—for example, trees constitute points. Before we learned to put points on paper, we could see that trees dotted the landscape. Thus points constitute the basic instrument of human thought, especially in mathematics from which all the sciences spring. The instants arise from the act of 'consciously' and 'purposely' moving from point to point, confirming the Russellian notion that time is 'relation between points'. Hence quantified time is human in origin, except that the internal sense of time (as duration of anything in the mind) must be recognised as making a psychological contribution to the invention of quantified time in that the external cycles used for quantified time (the years, for instance), have to have psychological anchors (meanings) which are the sense of duration of anything in the mind.

Secondly, in the absence of a standard time frame, what does it mean to claim that time intervals in a moving frame are shorter---shorter as against what kind of standard or universal time? What time intervals are they compared with since there is no standard time frame? (Note that you cannot say they are shorter as compared to other clocks outside the moving frame; that will bring in the Einstein theory of frames, as I will discuss presently.)

So we all, in the end, have to resort to using the clock or clocks based on the earth-year. Yet if we use the clocks then it is not correct to claim that time intervals in a moving frame are shorter; they are not naturally or normally (in its proper setting) shorter or longer; they are *normal* to that frame, or to its natural frame. The moving clock may only seem 'different' as *viewed* from the outside; but if that is the case then there is no puzzle.[125] The time of

---

[125] Otherwise it is difficult to see how the behaviour of one clock can affect all time, human physiology and even the material contents of atoms, e.g. muons. If time is defined as the passage of existence in consciousness, how can the behaviour of one clock affect it for all of us? There is still a lot of religious beliefs about time. Time dilation is one of them, so sweet to the religious in science because they can claim that "it is a unique mystery about time predicted by Einstein". In fact, it is not a mystery, let alone predicted by Einstein: he rather solved the little problem with his theory of frames—i.e. the dilated

the moving frame is not 'our' time; and it is not queer to its natural environment or setting. It is a strange phenomenon to those looking in from the outside, in breach of the Einstein theory of frames. In fact, it is irrelevant to anybody but those in the moving vehicle only.

The whole idea of studying other frames from the outside is fraught with difficulties; it can never be an exact science since the standard postulates that make our system work (and make it what it is) might be inapplicable outside our frame, or planet.[126] Speculations into other frames from our frame have been responsible for all the bizarre suppositions about time and space-time from mathematicians and cosmologists in general relativity. I don't think that kind of enterprise is justifiable, especially when it leads to theories that space-time may be infinite in its timelike directions. Space-time cannot be infinite because it is necessarily discrete---the year, for instance, is not infinite. It is only one; all other units of time derived from the year are also discrete and individual. The proper way to think of time as space-time is that its units are in perpetual procession (one year or second following another) to make time seem continuous; as such time can never be infinite.

Nothing illustrates the confusion about time in physics as a result of relativity and how it is misunderstood by scientists than the story of muons. By normal logic they should not last long enough to reach the earth; but they do. With the use of formulaic mathematics and concepts, physicists explain this by saying special relativity provides the answer as follows: the speed of muons is so

---

clock belongs to another frame to which it is running normally.

[126] I think one implication of this is that the laws of physics, or some of them, would differ from ours at least in some parts of the cosmos, if not all over. Einstein was really a very strange genius in physical thought. He introduced the notion of postulates for natural laws in frames. This idea may go very far indeed in the cosmos at large.

great that their internal clocks slow down. Using the theories of time dilation and the so-called twin paradox based on it, it is assumed that as the muons gathered speed and their internal clocks slowed down they aged less and thus are able to last long enough to reach the earth. To a logician or philosopher who understands relativity, this is so laughable as to choke him. It is really the best example of the confusion in physics about time in relativity. (1) Time dilation has nothing to do with the muons and how they behave, since time does not dilate internally. Lorentz found that a moving clock would be seen by outsiders as running slowly; but internally those carrying the moving clock would notice absolutely no difference in its performance. Einstein explained this with his theory of frames---the moving clock is in a different frame. There is no logical mechanism for this kind of episode to be able to control time *per se*. All other clocks would not run slower or faster; and since there is no such thing as an absolute time frame, or standard time, by which all other clocks can be compared, the moving clock's performance has no relevance at all in physics, because its carriers would notice no anomaly; and those outside who notice any anomaly should mind their own business since it is not their time. (2) The idea that muons have internal clocks is based on the Minkowski theory of space-time, where space and time are assumed to constitute one entity; and therefore the reasoning goes that, since the muons occupy space, and all space is space-time, they have their own internal clocks to keep or measure time for them. Again, any logician will describe this as nonsense; for after all, the Minkowski space is known to be fictitious and arbitrary with absolutely no logical validity.

    The basic idea in Time Dilation, which these writers rely on, is easily disproved thus: we know there are (roughly accurately) specific times by our normal clocks for the occurrences of certain events on this planet. Let us use Sunrise and Sunset for illustration. If Sunrise is usually 6 am, and Sunset is roughly 6 pm, as they are in some countries in the Tropics, it is inconceivable that a moving clock can force or influence these times to become 7. Am, and 7. Pm, on the planet all over just because one particular clock somewhere is running an hour late. "The dilation of time as a measure of moving clocks" can in no way influence all time *per se* on the planet.[127] It affects the performance of only one clock. Clocks are manufactured to reproduce specific time units, usually in seconds. If a particular clock, for whatever reason, is running erratically, there is no logical mechanism for its behaviour to affect all other clocks on the planet.

The reader will have noticed that the name of Lord Bertrand Russell comes up regularly in all discussions of relativity's interpretation. It is inevitable. Russell was highly respected by Einstein, and for very good reasons. He was the world's greatest philosopher at the time. He was also a great mathematician and logician of genius. A most attractive writer, who won the Nobel Prize for Literature, he wrote about every subject in philosophy, including novels to illustrate moral points. When relativity was announced, he abandoned many of his most cherished ideas as wrong without shame or even mild embarrassment. He was candid and honest in the most adorable way, completely dedicated to the truth no matter how it reflected on his own beliefs. Russell probably had no certain beliefs other than the pursuit of the truth

---

[127] In any case, the quandary was solved with Einstein's theory of frames, as previously mentioned. Referring to it as if it were some kind of strange metaphysical phenomenon we do not understand, is part of the religious reaction to many aspects of relativity and time. Einstein did not define time in metaphysics. He merely pointed out that it must be added to the study of phenomena as a distinct co-ordinate in its own right. He also noted that by the analysis of simultaneity it just cannot be possible that time is absolute that generally permeates the cosmos and the same everywhere in the Newtonian sense. The interpretations of these ideas were left to philosophers, who have noted that Time Dilation was resolved by Einstein when it led him to the discovery of special relativity to the effect that the universe consists of fragments, each with its own natural laws. Our natural physical laws are influenced by the two postulates he gave us, from which we learn that Time Dilation occurs in a different frame.

wherever it took him: via science, logic, mathematics or plain common sense, and linguistics. If he was certain that teaching mathematics to people from the cradle could save the world, he would have advocated that as his philosophy.

Concerning relativity specifically, in the later editions of his little book "Problems of Philosophy" he denounced his original philosophy as expressed in the book because of Einstein's theories, joking that whoever wrote the original ideas must have been a monkey, but nobody should suppose that the monkey looked, even remotely, like himself! No great philosopher has ever made such a confession; often associated with rulers, they all wrote imperious edicts as if they had discovered the final truth in logic and metaphysics.[128] Indeed, Russell later called his Fellowship dissertation "somewhat foolish" for the same reason, namely, the geometry used by Einstein had made his discussions of the foundations of geometry completely wrong, and he was happy to admit it and adopt the new Einstein theory. He wrote one of the best interpretations of relativity, still in use, under the title "ABC of Relativity". His book "The Analysis of Matter" can be divided into two. One section is about relativity; the other is mainly about his joint theory with A. N Whitehead to the effect that the world of

---

[128] No surprise, then, that Russell later put them in their deserved places (mostly of dishonour) in his monumental *History of Western Philosophy*. One complaint is that he never even once mentioned the name of Wittgenstein in this great book. The reason came from his contemporary, Sir Karl Popper---it was because, "In the long history of philosophy there are many more philosophical arguments of which I feel ashamed than philosophical arguments of which I am proud...Russell saw these things in that light, and so did I..." (From, *Modern British Philosophy*, By Bryan Magee, Secker & Warburg, London, 1971.) In 1959 Russell published his book, *My Philosophical Development*, in which he said he eventually had to reject Wittgenstein because he was talking 'logical mysticism' which was anathema to his basic nature. Of course he was right. Correctly defined, logical mysticism includes religion, mysticism and unscientific gibberish, all dressed-up to look like valid logical reasoning with a variety of linguistic trickery. Many aspects of philosophy in Oxford and Cambridge (and elsewhere) remain stuck in this kind of mud ever since.

sense is a construction, not an inference. Yet even this can be traced to relativity, since Einstein made man the observer part of the observed, meaning that man contributes something to the nature of physical reality---i.e. to help with the construction of that reality---and the book was published long after both special and general relativity. It is a moot point.

CONCLUSION

Time is so familiar that everybody thinks he or she knows what it is; in any case even those who desire to study it come with their own agenda of what it is or should be. Most of this knowledge of time has come from tradition and custom. To even try to sketch them will require a tome; there are an infinite number of ideas about time. Albert Einstein showed that even time in the clock is not what we take it to be. We owe this new theory of time not entirely to Einstein. The original idea or discovery that time is changeable came from the researches of the Dutch physicist Anton Lorentz. Previous to the new theory all mankind accepted what is generally referred to as "The Newtonian Absolute time." In fact, it wasn't Newton's idea alone; it was really a tradition whose origins are buried in religious tombs. The essential features of this time was that it's universal, generally covering the whole of the universe (because, of course), it's supposed to be divine in origin—and God was the "Creator", you may recall. As such it's also fixed; no human being dared to suggest that he or she could alter God's creation. Being fixed meant it was the same everywhere; that every unit of time here is the same everywhere. Logically, every schoolboy will now dispute this idea, for at least we know that the earth's orbit of the sun is what we call one year, our basic unit of time; but there're several bodies competing with our orbits, and we can guess that they take different periods to circle the sun, How can one unit of time based on the year be the same everywhere? Such questions influenced Einstein to make him declare that Lorentz's discovery of a different sort of time (from the traditional one) was rather a discovery of the real nature of time. This was not only amazing; it's revolutionary---even shocking---and freed mankind from the oppressive constraints of absolute time. Looking back, it's astonishing how slow knowledge of the external world tends to spread. The orbit of the sun is used for time and yet we could not connect that with the units of time we used and believed that they're independently created by God, were fixed and generally permeated the universe---how presumptuous. But then we have to remember that even our tiny earth was supposed to be the centre of the universe! Man may be small, but he has brains that seek to soar above the stars. And the clerics were the happiest for they exercised all this awesome power over the universe---how very presumptuous and vain, no wonder they invented the afterlife to come back after death.

Since his time, we have come to realise that everything we normally refer to as time (meaning every unit of time), is derived from all the planning we are able to do with the 24-hour periods of the revolutions of the earth, or the bigger 12-month round trip of the sun. Example are everywhere: time to catch a bus; time to go to the shop; time to go to school; time in sports; time for doing all the things we do. Another peculiarity is that we know time only in units. Some people believe that there is such a thing as 'silent time'. In sleep they claim that time is going silently. Well the clock is going anyway; but without the clock, is time going still? The positive answer is that the earth is moving us to different and new positions in the cosmos. We can extend this idea to chemistry.

Motion and chemistry can cause us 'a period of waiting'. That's time, of course, but how do we tell 'how much time it is'? I have explained this in the section entitled Time and Quantified Time. We quantify time into the specific units familiar to all of us. I argue that without quantification the word time has no precise meaning because it can be caused by anything---motion, chemistry, inertial, gravity, entropy, etc. These can all provide a period of waiting as time. What is culturally useful as time is to be found in the clock, but how did it get there? It is based on quantified time, any cycle at all counted as units of passing time.

The problem is that when we quantify time it becomes discrete. Yet all time has to be quantified (that is, reduced to units) to be useful in society, otherwise mentioning the word 'time' has no meaning except a private one to oneself. The old religious idea was that time existed all through the universe and we used our mathematics to create suitable time units for ourselves out of the blanket, universal entity. Once that notion was discredited, we had to find out how we get our time in units. But the process of quantification renders it discrete. Discrete time cannot run through nature, yet everybody refers to time as if it is running through nature. The old traditional view of time seems to me completely impossible to expunged from the human mind; we are wedded to a notion of existence linked with time as something like a thread running through nature and the mind can't free us of that. My opinion is that it is because once time is known it becomes part of existence, since one cannot define existence without the 'when' it's there; our nature is bound-up with the old definition; the new explanation may be more logical, but the old one is proving hard to expunge.

If we accept that we know time only in units, specific periods or moments, it implies that we also accept it as secular; for since the parameters for creating any inertial body's time units are different, every inertial body has got to invent its own time system. This is what has become known as 'Secular Time' in place of the old idea of time as "a fixed time system of divine origins covering the whole universe." Secular time can be logically traced from the parameters in our environment, meaning we create it here and will be applicable to us alone here in the universe. As stated by Einstein (quoted above): "There are as many times as there are inertial bodies". And the greatest philosopher of the time, our own Bertrand Russell, explained that our time is actually constructed as relation between points. For it should always be borne in mind that no one person, however clever, can decipher all the intricacies of time built up over so many centuries from before we're human to when we came down from the trees. So even Einstein required the help of the greatest philosopher alive.

Unfortunately, time is so closely associated with life that they don't seem to be separable; as a result we're happy (the mind is clinging to the old notion because it is 'happy') interpreting it as divine---but that was because life was also supposed to be divine. That God created life and the time to go with it. When that religious idea was rejected (or had to be abandoned in the face of scientific evidence: Darwin, astrophysics, scientific medicine, electricity, etc.), the Russellian query came in, showing how important philosophers are, namely, if cosmic time is abandoned, what really is measured by the clock? Part of my frustration is that nobody is interested in answering this question but rather wish to go on regarding time as if it is still running all through the cosmos and the same everywhere. Many university publishers reply to my submissions with the confession that they'd nobody on their staff with the necessary expertise to handle my material and wish me luck elsewhere. It's not that they can't see the importance of my work; the knowledge for assessing it is not there; we have reverted to taking time as it was before Einstein because we feel more comfortable with that.

However, frankly, not everybody accepts the secular theory of time. The religions have been completely defeated and they know it. They are pretending to be staunchly sticking to the divine origins of time because all of its mysteries cannot be laid bare by science; and they get away with this deceit because the whole of mankind is still behaving as if the new secular view of time has never been thought of at all. And they are able to do so mainly because time is reckoned in the same old manner we've become accustomed to over centuries, that is using the year as a unit of time and paring it down to the seconds. It is the basis of secular time; but it has been the foundation of religious time as well over several centuries. Is there a mystery here? I do not think so. We've always used the earth-year for time, hence our notions of 'years' to account for ageing. But the view was that we're merely sub-dividing the year with our mathematics to accord with divine time running through the universe and of which the year is part. It was not accepted as an original creation (construction as Russell put it) of time without divine intervention, which is what the new theory of secular time amounts to.

And we know that the new secular theory of time has to be true otherwise science wouldn't work; for this new assumption is proved by the number of quandaries that have become easily explicable on the secular theory of time derived as a construction based on "relation between points". Some of these issues that the secular theory of time allows us to formulate in consonance with science have been discussed in the book. They include the following: (1) That the past, present and future syndrome is memory of past events plus the consequences of these events carried with us to the present and whose remainder or results would be carried with us to the future---that history is seen as the march of events not time. People cause events, not time; the times are associated with the events to show the times and dates of occurrence; and obviously today's events have antecedents as they will have consequences. Einstein was therefore absolutely right in calling the syndrome an illusion. Of course it is true that time can cause events but only temporarily, or accidentally (like an avalanche going off after a while under the weight of additional snow), not as a continuous story. The continuous events of history (telling a human story), cannot be caused by time as a period of waiting. History is caused by sustained action deriving mostly from the past and often going on to cause the events of the future too. Time might even be irrelevant. For instance, soldiers in the trenches fighting over several days might not even notice that the old year had given way to a new one, and that Christmas had already gone.

(2) It is also easy to explain time travel on the secular theory of time; for the essential feature of that time is that it is discrete, not in a chain or a tread but proceeds unit by unit---exactly like the year. Time cannot be reckoned in any other way (this is one of my assertions and I'd stick my neck on it), for the real time is never known. What we call time is the number of units we create with any cyclical motion at all. Over the years, we've realised that the earth-year provides the most convenient cycle for reckoning time--- because of the astronomical features the seasons, the sun and its day and night system. But that should not mislead anybody about the nature of time. Orbiting the sun and calling each orbit a year and sub-dividing the year down to the seconds may be more convenient, but it's not much different from tapping your finger and counting them as the rate of the passage of time. Time is the repetitive cycles or motions we count as the rate of the passage of time. Real time in nature is either non-existent, in the absence of relevant parameters, or unknowable---unknowable not in the religious sense but in the sense that it's too complicated to decipher, involving several of the agents that can cause us to gain the sense of a period of waiting, which is what we call time.

So long as the year is seen as one unit of time that has to be repeated to continue, the time system based on the yearly cycle cannot be otherwise than discrete, proceeding unit by unit. In procession we get the continuity of time. I was rather surprised this was disputed by the very people who usually celebrate the end of the year so enthusiastically, so much that they deliberately engineered the birth of Jesus to occur at the very end of the year---crafty saints! You have to ponder it carefully to realise that it was a ruse; otherwise, as they wanted, you'd think it was real.[129] Meanwhile the Christian soldiers were marching on, the colonial exploiters were moving in, and the instruments of oppressions were being assembled. What is wrong in the world today started long ago in history and have gained such roots that the world will never be a peaceful place. I'm afraid we've lost all hope of a rational, well-organised, peaceful existence.

(3) Time travel has become popular in recent times; publishers are outbidding each other to publish books on time travel, because the man reputed to be the greatest logician ever to grace this planet, Kurt Godel, has claimed that his conversations with Einstein has convinced him that time travel is 'a scientific possibility'. On the other hand my books about secular time are never even touched.

---

[129] I thought it's real when I was young. The whole Christian myth worried me a great deal till I turned 23. Heaven, Hell, Day of Judgement and all the rest of it. My father was a school master and he pumped these Christian ideas into his children---or into children and adults generally. In those colonial days, teachers controlled the minds of small communities. Life was sad; that's how I look at it now. Here were adults living by the Christian ideology as bogus at it is, believing it to be true and that God is going to come down for the Day of Judgement. Then I came to Britain for further studies, though lacking even an ordinary elementary education. But I read widely till I came upon one woman writer in the small Psychologist Magazine of the early sixties. She it was who introduced Bertrand Russell to her readers, including yours truly, who devoured her every word. Luckily at the time Russell was imprisoned for ban the bomb marches and it was in the news every day. So I read his books and came across the sentence "That is why they invented Hell," and I lost all religious beliefs instantly. I didn't know that Hell was a literary invention---so must be Heaven, God and Satan, I concluded.

Nobody has ever asked to read any manuscript of mine. The majority of publishers and institutions never even reply to my submissions. But I am not bitter for time is a very serious subject and extremely difficult to write about. The human mind loves familiarity and often attacks new ideas fiercely. A theorist is lucky if he is not thrown to the lions as the initial reaction of the powerful to the unfamiliar; that they often turn out to have been correct but ahead of their times is immaterial. Human beings cannot act against the basic tendency ingrained in the brain. We progress but only slowly, any attempt to jump the queue could be dangerous. Better do the work and die off; the world will recognise it when the time comes, provided the work is good. If not, well, you tried. Ideas may incite strives and revolution, but not the theorist. He wouldn't even know the likely effects of what he writes. It is a defect in intellectuals and a welcome one: proposals should be advanced on logic and morality and left at that. What happens next must not on any account be predictable or predicted. It's more likely to go awfully wrong if predicted.

As I berate the greedy but shallow publishers of books on time travel, I also realise that they might be hedging their bets; and I honestly can't see anything wrong with that. For there is one mathematical theory that seems to make time travel possible. It is, as mentioned in the book, the 'great' Minkowski formula for equating space to time. He's not great; his theory too was not important either. But mathematicians seem not to care. They want time travel and the Minkowski proposal appears to make it feasible. He said he had achieved the union of space with time so that they've become one entity. I agree that if that is true then, given the curvature of space, time could go forward or backwards.

The reader does not have to take my word for it. Professor Arthur Eddington and Bertrand Russell have called the Minkowski theory arbitrary and fictitious. I'll settle for that. Since the

Minkowski formula is based on imaginary time coordinates, it certainly cannot make space take time with it when it curves. I condemn the theory as false, but he came close, very close. Unfortunately in science and logical thought coming close is not good enough---our lives depend on these ideas and we might lose them if they're not really true of the external world. Those people, invincibly arrogant people, who disparage philosophy should realise that science (electricity, running water, weather predictions, geostationary satellites, building and construction, agriculture, etc.) and scientific medicine (antibiotics. Immunology, haematology, all those shinning machines and scanners in our hospitals, blood pressure and its solutions, drugs and all the rest of it), have come from the numerous attempts to understand the world and nature generally for the benefit of mankind---the point is, they all arose as a result of arm-chair speculators imagining things sometime to the ridicule of their fellows. Science is only the material aspect of speculation; the really serious work is done by the philosophers thinking logically about everything privately. One unbroken rule is that all inquiries have to follow a logical route.

We accept that we do not know the truth about nature or the external world. I have mentioned the Platonic simile-of-the-cave, the logician Kurt Gödel's Incompleteness Theorem, the impossibility of knowing what time is except how time units are passing by, and the fact that we can never answer all the mysteries that surround time. We can never know everything. But all suggestions have to pass the test of logic because our lives depend on them and we do not want to walk into bottomless pits with our eyes wide open.

Over the centuries we've evolved various methods for establishing the reliable ideas by which we live. At the top of such methods is logic, the chief method for knowing what is reliable, safe or profitable. By such rules it is obligatory to reject the

Minkowski equation of space to time; and once that is rejected time travel is abolished. If his equation of space to time is not logically tenable, then time cannot travel (move, whirl, curve, gravitate in a black hole) together with space, moreover a time in a system of time reckoning that is strictly discrete, proceeding unit by unit.

(4) The passage of time has troubled thinkers down the centuries. The latest proposals centre on the arrow of time theory. It is assumed that time moves in one direction only; and that is regarded as a kind of arrow pointing to the direction of time. A thousand other suppositions tell us that this arrow can be reversed to allow time travel backwards; or it can have several dimensions; it can even turn worm-like and so forth

In fact, time does not move at all; all the movements in the story of history are physical and involve human motions and activities; they are what tell a story, and all history is a story of how human being have behaved or reacted to the environment. The year does not move from one year to two years. It is replicated to be years. So the arrow of time is not discussed at all in the book; my argument all through is that the new secular time is basically discrete---simply because the basic unit of time is only one; it does not stretch and it does not move; but it does expire completely. We start a new year at the end of the current one. This year replicates incessantly, hence the centuries. Discrete time passes by through the procession of its units, and that exactly is how the year passes by together with its fractions, from the second to the hours, weeks and months. The pulses of atomic time are included because they are based on the second. The radiation pulses of caesium 137 are used to measure the length of the second more precisely; that is all atomic time means; so you can't have atomic time without the year. The year remains the basic unit of time out of which all other units of time are derived with points or mathematics---always in association with the astronomical features of the earth and solar system. And the year is determinate; therefore the time system it affords us cannot be anything other than determinate or discrete time. Discrete time cannot march; it cannot spread; it cannot curve and it cannot be the same thing as space because it is created with points as applied to space, that is the reason it is discrete---from one point to another. Thus it is 'a product' of space, part of it but not the same thing. Time can still be called 'space-time' since we cannot have any unit of time without space. But space-time in the sense that space and time are unified into one entity as Minkowski proposed is obviously untenable in logic.

Thus, the concept of curved space-time by which time travel is said to be 'a scientific possibility' has been dismissed in the book as a bogey. Mathematicians hope it were true; it would make many things in the study of time simple. But it is based on imaginary time coordinates and therefore cannot be true in nature---only in mathematics.

(5) It seems to me that time is life and life is time because you cannot define life without reference to its 'when?' of existence; everything is set in time. We cannot dispute the theory that we live according to time or because there is time. What is mysterious or contentious is that we make the time. The religions have resisted the notion that time is secular precisely because of this implication. But the point is, to live is to be controlled completely by the earth's environment. This environmental conditions have been converted to time units, so that every moment of the earth's existence that has to support your own or anybody's life is time or a time unit, usually expressed in seconds, or even nanoseconds, each of which constitutes a portion of the earth's orbit round the sun; every moment is convertible to space round the sun. Worst of all (what the religions cannot accept), is the suggestion that this time is 'constructed' by man as Bertrand Russell put it. So what is life? To me it is time---the time allowing you to live on this planet.

I have also discussed other relevant topics such as Time Dilation, Gravity, Entropy and time, The clocks paradox and Twins paradox. From the point of view of secular time, these and many other mysteries can either be reasonably resolved or regarded as insoluble mysteries. We cannot explain everything; the life itself is not understood; we're orphans because we do not know whence we came, and should live carefully or cautiously without making demands on nature, since we're completely alien to it as it is also alien to us.

My final word is that nobody can ever discover the true nature of time, even if it exists. However, we have constructed a time system, according to Bertrand Russell. And it seems to me to raise two problems in philosophy: (1) Does time move or only the cycles we use for its reckoning do move deceptively as if it is time itself moving on? (2) Is that movement what we call "the passage of time", so that all man can ever know is the passage of time and therefore needs no theories to account for how time passes by? In any case the problem of the passage of time is solved.

All human thought and activities are based on, and regulated by, logic as the law of thought, since action is controlled by thought. The subject and predicate simple tool is also the basis of all thinking. Hence to me it's obvious that logically it is untenable to suppose that time slows down in a gravitational field, and even claim that the idea is one of Einstein's greatest discoveries, without knowing whether time moves at all, and how it moves.

In fact the questioning of time as absolute and fixed by God to cover the whole universe began when Einstein adopted the Lorentz discovery of local time, or t_1. After that Bertrand Russell made the debate about time really secular when he asked, 'what is measured by the clock if cosmic time is abandoned?' Nobody has felt able to answer this questions, namely why is time variable if it is fixed by

God, and what is measured by the clock if it is not fixed by God? Whether it speeds up or slows down here and there is irrelevant--- we need to know what it is, where it comes from, and whether it moves at all: the tasks attempted in this little book. It certainly cannot be the same as motion because time is known in units, and, as such, requires points or a digital existence, requiring human intervention (here we recall that Russell said it is constructed by man as relation between points---the years for instance.) Neither could it be interpreted as 'the totality of existence, or the passage of existence' because that would make it universal and local time could not be in existence; but Lorentz had discovered local time and so 'there is no longer a universal time'.

Let's come down from the grand theory to mundane simplicity. Life is not theoretical; we made it so, mostly to render the world habitable and emotionally palatable. The mundane justification of secular time, as opposed to divine time full of cosmic mysteries (a justification in logic, linguistics, mathematics and all, but religion) is this: life depends on what exists or happens in astronomy. Therefore, so long as there is only one day and one year in all astronomy there can be no such thing as the passage of time. The days as the passage of time is a myth; and the years, too, as the passage of the centuries is false. The days are caused by the revolutions of the earth across the face or the 'constant' glare of the sun; and of course, we all know that the year ends and restart ever 31st December. Yet these are what we regard as the passage of time—for our personal ages, for instance! Even worse, for the age of the universe.

I conclude that(1) There are no days and years in nature the passage of which can demonstrate the passage of time. (2) The passage of time is an illusion caused by the repetitive cycles we employ to indicate the passage of time when we thought that time was passing to Judgement Day all through nature like a thread. We know now that it is not true. Time moves through the motions of the cycles---the years for instance---that we use to tell as the passage of time. The clock is used to breakdown the long year into manageable units for cultural use. The years are mere physical cycles we count to give us arithmetically satisfying numbers of such cycles we have seen go pass our lives, a false evidence of longevity. In reality we know nothing about longevity; the same applies to using the year as the yardstick for measuring the cosmos in age, distance and everything! That's how pathetic we are---worth nothing, knows nothing for sure and going nowhere. The religions are right: they give us hope, a false hope, yes, but better than nothing whatsoever.

REFERENCES

ALBERT EINSTEIN (1879-1955) ---SPACE TIME, an article in the 1926/27 (13th) edition of the Encyclopaedia Britannica. Also, RELATIVITY, in the same edition.

---NATURE No. 106, 782, (1921), almost the whole issue was devoted to the confirmation of Einstein's new theory of gravity.

---*The Meaning of Relativity*, Princeton University Press, 1966.

---*The Evolution of Physics*, (With Leopold Infeld) Cambridge 1838.

---RELATIVITY, Routledge Classics, London and New York, 2001.

HERMANN MINKOWSKI (1864-1909) ---He first mentioned his supposition in a lecture in cologne, known as *Raum und Zeit* (Space and Time) Cologne 21st September, 1908.

--- Herman Minkowski AdP 47, 927 (1915)

---Herman Minkowski, Goett. Nachr., 1908 p53. Reprinted in Gesammelte Abhandlungen von Herman Minkowski. Vol. 2, p352. Teubner, Leipzig 1911.

BERTRAND RUSSELL, FRS (1872-1970)---*Our Knowledge of the External World*, George Allen & Unwin, 1922.

--- *Mysticism & Logic*, George Allen & Unwin, 1976: a collection of important essays first published in 1917.

---ABC OF RELATIVITY, George Allen & Unwin, 1958 (recently revised by Professor Felix Pirani---first published in 1925.

---*History of Western Philosophy*, George Allen & Unwin, 1946.

---*My Philosophical Development*, George Allen & Unwin, 1958.

---*The Analysis of Matter*, George Allen & Unwin, 1927.

MORRIS KLINE: *Mathematics in Western Culture*, Allen & Unwin, London, 1954.

SIR ARTHUR STANLEY EDDINGTON, FRS (1862-1944)

---*The Expanding universe*, University of Michigan Press, Ann Arbor, 1933

---*The Combination of Relativity Theory and Quantum Theory*, Communication of the Dublin Institute of Advanced Studies, Dublin, 1943.

---*The Mathematical Theory of Relativity*, Cambridge, second ed. 1930.

---*The Nature of the Physical World*, Ann Arbor, Michigan, 1958.

---Philosophy of Physical Science, Cambridge, 1949.

---The Theory of Relativity and its Influence on Scientific Thought, Oxford, 1922.

---Space, Time and Gravitation, Cambridge, 1920.

SIR JAMES JEANS, FRS: *Physics and Philosophy*, Cambridge, 1942.

---*The Mysterious Universe*, Cambridge, 1930.

---The New Background of Science, Cambridge, 1933.

PROFESSOR A.N. WHITEHEAD: *The Concept of Nature*, Ann Arbor, Michigan, 1957.

---Science and the Modern World, Cambridge, 1922.

---An Inquiry Concerning the Principle of Natural Knowledge, Cambridge, 1919.

---*Nature and Life*, Cambridge, 1934.

---Process and Reality: An Essay in Cosmology, Cambridge, 1929.

---*Essays in Science and Philosophy*, Rider & Co., London, 1948.

---The Principle of Relativity, Cambridge, 1922.

Professor BANESH HOFFMANN: *The strange Story of the Quantum*, Dover Pub. Inc. New York, 1959.

Professor STEVEN F. SAVITT (ed.) *Times Arrows Today: Recent Physical and Philosophical Work on the Direction of Time*, Cambridge, 1995.

CHARLES A. FRITZ: Bertrand Russell's Construction of the External World, Routledge & Kegan Paul, London, 1952.

Professor JEREMY BERNSTEIN: Albert Einstein and the Frontiers of Physics, Oxford, 1996.

Professor RICHARD FEYNMAN: Lectures---*The Character of Physical law*. MIT Press, 1967. There are several volumes of the Feynman lectures and they are all worthy of serious study.

Abraham Pais, "Subtle is The Lord: The Life and Science of Albert Einstein", Oxford, 1982. Professor Pais has methodically provided details of almost all the original papers relevant to relativity. His list is so exhaustive I don't know of a better one anywhere.

WHAT REMAINS TO BE DISCOVERED, By Sir John Maddox, A Touchstone Book, Simon & Schuster, 1999.

INDEX

3+132, 45, 67, 72, 100, 106, 121, 156, 159, 178
31st............49, 129, 138, 200
4-D29, 30, 45, 53, 57, 68, 110, 118, 120, 121, 127, 145, 163, 173, 176
AD.....................117
Afterlife.....................160
agriculture 196
alien.........................199
Alps..............................7
ambiguity 39
American.....................15
Appendix...............123, 179
Appendixes..................23
Archbishop............44, 57
argument18, 34, 168, 181, 197

Aristotle8, 23, 40, 96, 115, 125
arithmetic6, 42, 49, 50, 74, 92, 93, 116, 158
Arthur6, 10, 17, 30, 40, 43, 118, 163, 166, 170, 173, 195
article............23, 29, 146, 203
artificial45, 49, 61, 68, 69, 72, 163
a-search-and-rescue............51
asteroids 86, 96
astronomers 32, 148, 171, 173
astronomical28, 31, 35, 38, 43, 49, 51, 108, 124, 145, 162, 193, 197
astronomy21, 22, 24, 26, 52, 78, 114, 122, 129, 161, 200

astrophysics 6, 8, 118, 146, 159, 191
atom 81, 84
Bergson 114, 115, 135
Bernstein 29, 58, 72, 150, 171
Bertrand 6, 11, 19, 23, 28, 29, 31, 32, 40, 44, 52, 60, 61, 62, 63, 64, 68, 71, 77, 95, 96, 114, 115, 117, 124, 127, 146, 156, 159, 161, 170, 172, 173, 174, 175, 178, 179, 180, 185, 191, 194, 195, 198, 199, 205
Bible 149
biblical 117
biology 143, 178
biophysics 83, 146
Branson 95, 140
Britain 54, 166, 194
Britannica 63, 203
caesium 197
Cambridge 19, 20, 21, 40, 41, 68, 186, 203, 204, 205
cave 53, 165
CERN 110
cesium 56, 181
chicken-and-egg 48
Christian 194
Christians 93
Circadian 142
civilisation 25
clarification 52

CLARIFICATIONS 25
clockmaker 164
clock-makers 162, 164
clocks 57, 60, 122, 133, 136, 137, 138, 139, 140, 141, 142, 143, 144, 151, 172, 181, 182, 184, 199
Cogito 115
Cologne 57, 203
colonial 21, 194
Copernicus 39
cosmologist 95
counterintuitive 49
creation 30, 45, 65, 73, 75, 91, 93, 102, 122, 156, 164, 188, 192
creation/construction 30
CUP 9
cycle 13, 15, 46, 48, 49, 50, 56, 68, 78, 83, 87, 108, 109, 117, 129, 137, 145, 151, 152, 154, 157, 160, 164, 190, 193, 194
cycles 7, 8, 11, 15, 21, 28, 36, 37, 38, 39, 46, 48, 49, 50, 51, 52, 56, 77, 79, 83, 86, 92, 97, 99, 101, 106, 109, 113, 128, 130, 131, 145, 151, 152, 153, 154, 155, 157, 159, 160, 161, 162, 163, 164, 175, 179, 182, 193, 199, 200

Darwin........83, 131, 178, 191
date.....................................44
David 44, 67, 70, 73, 121, 163
Dawn..........................42, 122
Day 21, 27, 51, 55, 93, 106, 109, 132, 138, 143, 194, 200
Day&Night..................51, 143
Daylight.............................21
days 9, 12, 13, 19, 21, 24, 25, 27, 28, 33, 34, 36, 38, 43, 50, 57, 78, 92, 98, 103, 108, 119, 124, 129, 193, 194, 200
December ...49, 129, 138, 200
decomposed........................96
deduce 81, 125
deduction...........................37
deductive 62, 179
delayed-reaction.......145, 165
Denoting..........................115
Descartes..................115, 116
dimensional 44, 60
dimensions 127, 197
Dingle.......................146, 178
doctrine................14, 81, 168
doctrines 114, 167
dogmatic..........................138
ds2 174
Dualism 116
dubious 148

due 5, 10, 24, 26, 33, 52, 57, 62, 70, 75, 90, 106, 108, 116, 119, 136, 146, 152, 181
Duration 12, 33, 35, 48, 51
durations...........................47
duration—the 50
during 11, 22, 27, 33, 34, 48, 55, 129
Dutch..........................26, 188
dynamic......39, 171, 172, 174
Eddington 6, 10, 17, 19, 20, 24, 26, 28, 30, 40, 43, 44, 45, 53, 65, 68, 71, 72, 76, 100, 106, 107, 111, 112, 118, 129, 134, 136, 144, 145, 159, 163, 166, 170, 173, 174, 176, 179, 195
edicts 186
edition..................17, 23, 203
editions 186
editor 23, 144, 152
Einstein 6, 8, 9, 10, 12, 14, 18, 19, 20, 23, 24, 25, 26, 27, 29, 30, 32, 36, 37, 39, 40, 41, 42, 43, 44, 45, 46, 47, 54, 57, 58, 60, 61, 63, 65, 66, 67, 68, 69, 71, 72, 73, 75, 76, 77, 78, 79, 81, 84, 86, 89, 95, 96, 97, 98, 99, 104, 105, 106, 107, 109, 110, 111,颠112, 113, 114,

116, 117, 118, 119, 120, 121, 122, 123, 125, 126, 130, 131, 133, 134, 135, 136, 137, 139, 140, 147, 148, 150, 152, 156, 158, 159, 160, 161, 163, 169, 170, 171, 172, 173, 174, 175, 176, 177, 178, 179, 180, 181, 182, 183, 184, 185, 186, 188, 191, 193, 194, 199, 203, 205, 206
Einsteinian 37, 122
electricity 86, 126, 191, 196
electrodynamics 14
electromagnetic 150
electronic 108
electrons 13
Encyclopaedia 63, 203
Encyclopedia 45
entity 7, 27, 38, 40, 41, 42, 44, 66, 68, 69, 73, 76, 77, 78, 93, 98, 100, 106, 110, 113, 115, 119, 120, 121, 133, 141, 144, 149, 152, 159, 160, 168, 169, 173, 176, 184, 190, 195, 198
entropy 93, 132, 133, 135, 144, 148, 152, 190
ephemeral 84
equating 107, 160, 195

equation 30, 36, 47, 60, 62, 64, 69, 73, 75, 107, 110, 121, 156, 159, 170, 197
Esynched 60
Fellowship 186
Feynman 13, 14, 38, 77, 101, 153, 206
Feynman: 14
fiction .63, 107, 115, 163, 170
first-class 115
four-dimensional 29, 44, 57, 67, 68, 69, 73, 74, 75, 106, 120, 121, 147
Frontiers 29, 58, 150, 171, 205
Future 9
galaxies 17
Gates 110
genius 23, 40, 43, 63, 67, 90, 121, 124, 126, 135, 163, 180, 183, 185
geniuses 25
geometrical .29, 58, 70, 71, 72
geometry 29, 30, 44, 45, 53, 57, 67, 68, 70, 110, 118, 120, 121, 127, 163, 173, 176, 186
geostationary 196
God 12, 15, 17, 20, 25, 32, 37, 53, 81, 82, 83, 90, 94, 111, 113, 115, 116, 117, 119,

124, 128, 131, 137, 141, 155, 188, 191, 194, 199
Godel116, 119, 122, 134, 160, 194
Gödel 99, 116, 196
Gods 94, 127
Gold 112
Gottfried 44, 50
Gottingen 44, 67
gravitational 94, 199
gravity23, 25, 32, 46, 53, 58, 70, 71, 73, 82, 87, 102, 105, 125, 132, 133, 135, 149, 150, 151, 166, 171, 190, 203
Gribbin 29, 58, 60, 135
Guardian 23
guns 147
gurus 20
habit 11, 58, 75, 80
half-wits 30
Harvard 18, 20
Heaven 94, 148, 194
heavy 147
height 178
Henceforth 57, 168
Henri 114, 115
Henry 114
Herbert 146
highbrow 12
highest 14, 40, 65, 134, 139

Hilbert44, 67, 70, 73, 121, 163
History20, 107, 114, 186, 193, 204
homogeneous 179
ict30, 69, 73, 110
Idealism 13, 15, 29
incisive 84, 146, 165
Incompleteness 116, 196
inertia34, 48, 81, 88, 102, 145, 165
infinite19, 23, 26, 37, 53, 60, 63, 65, 66, 73, 84, 117, 183, 188
infinity 17, 34, 35, 73, 76
infirm 111
institution 19, 21
intuition 157
Invariance 32
Irish 116
Isaac 117
James 116
Jesus 194
John 29, 58, 144, 206
Kafka 120
Kant 114, 135
Karl 75, 98, 146, 186
Kepler 156
kinematic 40
kinetics 165
Kurt99, 116, 120, 122, 160, 194, 196

KWASI 24
laboratories 180
LASER 101
Leibniz 44, 50, 124, 158
Local 57
locality 66, 139, 172
logic 5, 7, 8, 10, 19, 31, 35, 38, 39, 46, 47, 56, 58, 66, 69, 74, 75, 77, 78, 83, 89, 97, 99, 114, 119, 123, 127, 128, 132, 139, 141, 153, 154, 156, 160, 166, 167, 174, 179, 180, 181, 183, 186, 195, 196, 198, 199, 200
logician 15, 60, 124, 159, 184, 185, 194, 196
LONDON 24
Lorentz 6, 11, 22, 26, 30, 31, 37, 46, 57, 60, 66, 70, 74, 86, 89, 98, 109, 118, 126, 131, 136, 138, 161, 169, 171, 172, 174, 181, 184, 188, 199
Lorentz/Einstein 118
Lorentz-Einstein 22
Macro 63
Maddox 144, 206
magicians 142
Mathematical 18, 45, 54, 57, 100, 111, 166, 170, 173, 176, 179, 204

mathematically 6, 32, 86, 92, 123, 161
mathematician 20, 43, 60, 65, 71, 73, 86, 118, 124, 127, 167, 168, 173, 185
mechanics 36, 39, 50, 56, 78, 85, 87, 136, 143, 162
mechanism 12, 48, 49, 59, 96, 138, 141, 179, 184, 185
Medal 112
medium 150
memory 13, 19, 26, 64, 107, 179, 192
metaphysical 12, 87, 137, 139, 178, 185
metaphysically 6, 162
metaphysics 19, 21, 86, 114, 142, 167, 185, 186
mid-day 49
midnight 49, 138
milieu 53
Milky 32
mind 7, 8, 9, 13, 34, 36, 45, 46, 48, 53, 56, 61, 65, 78, 83, 90, 92, 93, 101, 103, 105, 110, 111, 115, 118, 135, 160, 162, 175, 179, 182, 184, 190, 191, 195
minds 35, 56, 95, 98, 122, 125, 138, 194
Minkowski 29, 30, 32, 37, 41, 44, 45, 53, 54, 57, 58, 60,

63, 64, 67, 68, 69, 70, 71, 72, 73, 75, 76, 77, 78, 106, 110, 115, 118, 120, 121, 127, 130, 135, 141, 142, 144, 145, 147, 151, 156, 159, 160, 163, 168, 169, 170, 173, 174, 176, 177, 178, 184, 195, 197, 198, 203, 204
mischief-makers 124
misconception 37
MISCONCEPTIONS 177
MIT 20, 206
momentum 43, 77, 165
months 24, 28, 43, 83, 85, 92, 103, 124, 150, 154, 158, 197
moon 100, 142, 161, 162
Moore 14
moral 185
morality 195
Newton 18, 75, 82, 117, 122, 125, 138, 163, 188
Newtonian 22, 57, 118, 139, 171, 185, 188
nightfall 49, 164
Nobel 13, 14, 23, 26, 53, 82, 180, 185
non-interacting 27, 41, 68, 77, 110, 150
non-professional 22
ocean 149

off-chance 18, 40, 111
over-reach 21
Oxbridge 21, 100, 146
Oxford 20, 21, 29, 40, 58, 186, 205, 206
perspectives 10, 114, 125
PhDs 60
Philosopher/Scientist 32, 65
Physicist 26
physicists 61, 70, 76, 112, 177, 183
physics 14, 18, 19, 20, 24, 26, 34, 35, 45, 54, 56, 57, 61, 62, 63, 68, 71, 72, 74, 75, 76, 78, 86, 101, 106, 112, 114, 115, 125, 126, 141, 142, 146, 163, 168, 169, 174, 177, 179, 180, 183
physio/chemical 28
physio/chemistry 126, 129
Planck 163
post-relativity 108, 158
postulate 127
postulates 9, 67, 125, 127, 183, 185
premise 70, 110, 118
pre-scientific 134
Princeton 18, 20, 101, 203
Principle 41, 68, 76, 110, 129, 176, 205
psychological-time 112

psychology 12, 28, 35, 114, 115, 121
Pythagoras 44, 64, 75, 85, 92, 115
Pythagorean......116, 138, 174
QED 13, 18, 35, 37, 48, 52, 56, 88, 101, 102
qualitative/physical 167
qualitatively......................168
quandaries 26, 192
quandary 13, 15, 108, 132, 156, 185
quanta...........................13, 82
quantification 47, 106, 123, 159, 190
quantified 56, 153, 158, 159, 161, 164, 179, 181, 190
quantum 13, 15, 18, 37, 52, 55, 56, 67, 81, 85, 87, 94, 95, 100, 101, 102, 105, 114, 181
Raum.........................57, 203
recur 37
References........................102
Regular..............................47
relativity 15, 18, 27, 28, 29, 30, 31, 32, 33, 40, 41, 42, 44, 45, 46, 53, 57, 58, 60, 61, 62, 63, 64, 65, 67, 68, 70, 71, 72, 76, 78, 81, 84, 86, 95, 97, 100, 105, 106, 115, 118, 119, 120, 121, 125, 126, 130, 134, 135, 136, 144, 147, 150, 152, 160, 163, 169, 170, 171, 172, 173, 174, 176, 177, 179, 183, 185, 186, 206
replicate 82, 109, 129
replicated......50, 91, 132, 197
Returning.........................116
re-use................................23
revolution 37, 75, 126, 195
revolutionary 32, 61, 71, 114, 139, 178, 189
Richard 13, 38, 77, 95, 140, 153
Russell 6, 7, 11, 12, 13, 14, 18, 19, 20, 21, 23, 24, 26, 28, 29, 30, 31, 32, 37, 39, 40, 41, 44, 45, 46, 48, 50, 51, 52, 53, 58, 60, 61, 63, 64, 68, 71, 72, 74, 75, 76, 77, 82, 92, 95, 96, 98, 103, 106, 114, 115, 117, 119, 124, 125, 127, 128, 146, 156, 159, 161,颠163, 168, 170, 172, 173, 174, 175, 176, 178, 179, 180, 185, 186, 191, 192, 194, 195, 198, 199, 205
Russellian 66, 84, 160, 182, 191
Rutherford 180
S=CT 156

scholars 14, 18, 21, 26, 37, 48, 76, 127, 135, 168
scholarship 8
secular 8, 9, 12, 15, 18, 25, 28, 37, 40, 42, 45, 78, 82, 84, 89, 98, 110, 113, 114, 123, 124, 125, 134, 150, 157, 160, 174, 190, 192, 193, 194, 197, 198, 199, 200
sermons 89, 100, 109, 132, 138, 144
simile .. 53, 156, 159, 166, 196
simile-of-the-cave 156, 159, 166, 196
simple 10, 11, 14, 22, 33, 66, 70, 78, 112, 126, 130, 131, 139, 141, 145, 146, 159, 172, 198, 199
simultaneity 114, 185
single-handed 180
sketched 42, 62
Space 26, 32, 41, 57, 60, 82, 100, 158, 183, 203, 205
SPACE-TIME 171
space-timed 175
speculators 196
States 43
stellar 88
sub-atomic 100, 105
subdivide 38
sub-divide 155
sub-divide 161

subject 5, 12, 19, 22, 26, 29, 31, 45, 69, 101, 114, 116, 129, 173, 179, 180, 185, 195, 199
subjectivity 159
subject-matter 180
sundial 33, 35, 49
sunlight 21
Sunrise 184
Sunset 184
temporal 28, 35, 79, 130
Theorem 116, 196
thermodynamics 148
thesis 119
Thomases 43
time-controlled 143
time-dependent 56
timelike 63, 183
times 6, 18, 22, 26, 32, 39, 40, 46, 51, 83, 91, 107, 108, 111, 114, 123, 129, 140, 160, 169, 173, 174, 184, 191, 192, 194
time-span 149
time-system 41
time-value 70
Traditional 19
Whitehead 6, 12, 13, 14, 15, 20, 21, 28, 36, 37, 41, 44, 50, 53, 65, 68, 75, 76, 106, 109, 115, 124, 129, 150, 163, 170, 176, 187

world-class 21	Yourgrau 43, 44, 68, 70, 99, 119, 121
world-famous 135	
world-shaking....................116	Zeit 57, 203
yesterday 9, 13, 21	Zero 42, 111, 117

www.ingramcontent.com/pod-product-compliance
Lightning Source LLC
Chambersburg PA
CBHW070228180526
45158CB00001BA/188